Congressional
Research
Service

Geoengineering:
Governance and Technology Policy

Kelsi Bracmort
Specialist in Agricultural Conservation and Natural Resources Policy

Richard K. Lattanzio
Analyst in Environmental Policy

January 2, 2013

Congressional Research Service

7-5700

www.crs.gov

R41371

CRS Report for Congress

Summary

Climate change policies at both the national and international levels have traditionally focused on measures to mitigate greenhouse gas (GHG) emissions and to adapt to the actual or anticipated impacts of changes in the climate. As a participant in several international agreements on climate change, the United States has joined with other nations to express concern about climate change. However, in the absence of a national climate change policy, some recent technological advances and hypotheses, generally referred to as "geoengineering" technologies, have created alternatives to these traditional approaches. If deployed, these new technologies could modify the Earth's climate on a large scale. Moreover, these new technologies may become available to foreign governments and entities in the private sector to use unilaterally—without authorization from the United States government or an international treaty—as was done in the summer of 2012 when an American citizen conducted an ocean fertilization experiment off the coast of Canada.

The term "geoengineering" describes this array of technologies that aim, through large-scale and deliberate modifications of the Earth's energy balance, to reduce temperatures and counteract anthropogenic climate change. Most of these technologies are at the conceptual and research stages, and their effectiveness at reducing global temperatures has yet to be proven. Moreover, very few studies have been published that document the cost, environmental effects, socio-political impacts, and legal implications of geoengineering. If geoengineering technologies were to be deployed, they are expected to have the potential to cause significant transboundary effects.

In general, geoengineering technologies are categorized as either a carbon dioxide removal (CDR) method or a solar radiation management (SRM) method. CDR methods address the warming effects of greenhouse gases by removing carbon dioxide (CO_2) from the atmosphere. CDR methods include ocean fertilization, and carbon capture and sequestration. SRM methods address climate change by increasing the reflectivity of the Earth's atmosphere or surface. Aerosol injection and space-based reflectors are examples of SRM methods. SRM methods do not remove greenhouse gases from the atmosphere, but can be deployed faster with relatively immediate global cooling results compared to CDR methods.

To date, there is limited federal involvement in, or oversight of, geoengineering. However, some states as well as some federal agencies, notably the Environmental Protection Agency, Department of Energy, Department of Agriculture, and the Department of Defense, have taken actions related to geoengineering research or projects. At the international level, there is no international agreement or organization governing the full spectrum of possible geoengineering activities. Nevertheless, provisions of many international agreements, including those relating to climate change, maritime pollution, and air pollution, would likely inform the types of geoengineering activities that state parties to these agreements might choose to pursue. In 2010, the Convention on Biological Diversity adopted provisions calling for member parties to abstain from geoengineering unless the parties have fully considered the risks and impacts of those activities on biodiversity.

With the possibility that geoengineering technologies may be developed and that climate change will remain an issue of global concern, policymakers may determine whether geoengineering warrants attention at either the federal or international level. If so, policymakers will also need to consider whether geoengineering can be effectively addressed by amendments to existing laws and international agreements or, alternatively, whether new laws and international treaties would need to be developed.

Contents

Introduction 1

Geoengineering Governance 3
    Risk Factors 4
    Policy Considerations 5
Geoengineering Technologies 9
    Carbon Dioxide Removal 10
        Carbon Capture and Sequestration 10
        Ocean Fertilization 12
        Afforestation 13
        Enhanced Weathering 14
    Solar Radiation Management 15
        Enhanced Albedo (Surface and Cloud) 16
        Aerosol Injection 18
        Space-Based Reflectors 19
The Debate over the Methods of Oversight 20
    The Status Quo 20
    Threshold for Oversight 20
    Methods for Oversight 21
    Moratoriums 22
The Debate over Oversight and Governmental Involvement 23
    State Policies Addressing Geoengineering 23
    National Policies Addressing Geoengineering 24
        Current U.S. Policies Addressing Geoengineering 24
        Potential Roles for Federal Agencies and Other Federally Funded Entities 25
    International Cooperation on Geoengineering 29
Conclusion 38

Figures

Figure 1. Geoengineering Technology Options 9
Figure 2. Cloud Whitening Schematic 18

Tables

Table 1. Scientific Underpinnings for Different Perspectives on Geoengineering 7
Table 2. Six Types of Functions Federal Entities Can Perform and Selected Federal
    Entities Authorized to Perform Them 27

Contacts

Author Contact Information 39

Introduction

Climate change has received considerable policy attention in the past several years both internationally and within the United States.[1] A major report released by the Intergovernmental Panel on Climate Change (IPCC) in 2007 found widespread evidence of climate warming, and many are concerned that climate change may be severe and rapid with potentially catastrophic consequences for humans and the functioning of ecosystems.[2] The National Academies maintains that the climate change challenge is unlikely to be solved with any single strategy or by the people of any single country.[3]

Policy efforts to address climate change use a variety of methods, frequently including mitigation and adaptation.[4] Mitigation is the reduction of the principal greenhouse gas (GHG) carbon dioxide (CO_2) and other GHGs.[5] Carbon dioxide is the dominant greenhouse gas emitted naturally through the carbon cycle and through human activities like the burning of fossil fuels. Other commonly discussed GHGs include methane, nitrous oxide, hydroflourocarbons, perflourocarbons, and sulfur hexaflouride. Adaptation seeks to improve an individual's or institution's ability to cope with or avoid harmful impacts of climate change, and to take advantage of potential beneficial ones.[6]

Some observers are concerned that current mitigation and adaptation strategies may not prevent change quickly enough to avoid extreme climate disruptions. Geoengineering has been suggested by some as a timely additional method to mitigation and adaptation that could be included in climate change policy efforts. Geoengineering technologies, applied to the climate, aim to achieve large-scale and deliberate modifications of the Earth's energy balance in order to reduce temperatures and counteract anthropogenic (i.e., human-made) climate change; these climate modifications would not be limited by country boundaries. As an unproven concept, geoengineering raises substantial environmental and ethical concerns for some observers.[7] Others

---

[1] For more information on the policy issues associated with climate change, see CRS Report RL34513, *Climate Change: Current Issues and Policy Tools*, by Jane A. Leggett; and CRS Report R40643, *Greenhouse Gas Legislation: Summary and Analysis of H.R. 2454 as Passed by the House of Representatives*, coordinated by Mark Holt and Gene Whitney.

[2] See IPCC website at http://www.ipcc.ch/publications_and_data/ar4/syr/en/contents.html; and United Nations Environment Programme, *Climate Change Science Compendium 2009*, 2009, http://www.unep.org/pdf/ccScienceCompendium2009/cc_ScienceCompendium2009_full_highres_en.pdf.

[3] The National Academies, *Ecological Impacts of Climate Change*, 2009, http://dels-old.nas.edu/dels/rpt_briefs/ecological_impacts.pdf.

[4] H.R. 2454, the American Clean Energy and Security Act of 2009 (Waxman/Markey), and S. 1733, the Clean Energy Jobs and American Power Act (Kerry/Boxer), were the primary energy and climate change legislative vehicles in the 111th Congress. For a comparison of key greenhouse gas emission control provisions in both the House and Senate, see CRS Report R40556, *Market-Based Greenhouse Gas Control: Selected Proposals in the 111th Congress*, by Larry Parker, Brent D. Yacobucci, and Jonathan L. Ramseur.

[5] For more information on proposed climate change mitigation, see CRS Report R40236, *Estimates of Carbon Mitigation Potential from Agricultural and Forestry Activities*, by Renée Johnson, Jonathan L. Ramseur, and Ross W. Gorte.

[6] For more information on proposed climate change adaptation measures, see CRS Report R40911, *Comparison of Climate Change Adaptation Provisions in S. 1733 and H.R. 2454*, by Eugene H. Buck et al.

[7] Alan Robock, "20 Reasons Why Geoengineering May Be a Bad Idea," Bulletin of the Atomic Scientists, May/June 2008.

---

respond that the uncertainties of geoengineering may only be resolved through further scientific and technical examination.[8]

Proposed geoengineering technologies vary greatly in terms of their technological characteristics and possible consequences. They are generally classified in two main groups:

- *Solar radiation management (SRM) method:* technologies that would increase the reflectivity, or albedo, of the Earth's atmosphere or surface, and

- *Carbon dioxide removal (CDR) method:* technologies or practices that would remove CO_2 and other GHGs from the atmosphere.

Much of the geoengineering technology discussion centers on SRM methods (e.g., enhanced albedo, aerosol injection). SRM methods could be deployed relatively quickly if necessary, and their impact on the climate would be more immediate than that of CDR methods. Because SRM methods do not reduce GHG from the atmosphere, global warming could resume at a rapid pace if a deployed SRM method fails or is terminated at any time. At least one relatively simple SRM method is already being deployed with government assistance.[9] Other proposed SRM methods are at the conceptualization stage. CDR methods include afforestation, ocean fertilization, and the use of biomass to capture and store carbon.

The 112th Congress did not take any legislative action on geoengineering. In 2009, the House Science and Technology Committee of the 111th Congress held hearings on geoengineering that examined the "potential environmental risks and benefits of various proposals, associated domestic and international governance issues, evaluation mechanisms and criteria, research and development (R&D) needs, and economic rationales supporting the deployment of geoengineering activities."[10] Some foreign governments, including the United Kingdom's, as well as scientists from Germany and India, have begun considering engaging in the research or deployment of geoengineering technologies because of concern over the slow progress of emissions reductions, the uncertainties of climate sensitivity, the possible existence of climate thresholds (or "tipping points"), and the political, social, and economic impact of pursuing aggressive GHG mitigation strategies.[11]

Congressional interest in geoengineering has focused primarily on whether geoengineering is a realistic, effective, and appropriate tool for the United States to use to address climate change.

---

[8] Jamais Cascio, "It's Time to Cool the Planet," *The Wall Street Journal*, June 15, 2009; and American Meteorological Society, "Proposals to Geoengineer Climate Require More Research," press release, July 21, 2009, http://www.ametsoc.org/amsnews/2009geoengineering.pdf.

[9] Enhanced albedo is one SRM effort currently being undertaken by the U.S. Environmental Protection Agency. See the Enhanced Albedo section below for more information.

[10] U.S. Congress, House Committee on Science and Technology, *Geoengineering: Assessing the Implications of Large-Scale Climate Intervention*, 111th Cong., 1st sess., November 5, 2009.

[11] A tipping point is defined as a critical threshold at which a tiny perturbation can qualitatively alter the state or development of a system. For more discussion on climate sensitivity, thresholds, and other scientific concerns, please see CRS Report RL34266, *Climate Change: Science Highlights*, by Jane A. Leggett. For international obligations pertaining to climate change, see CRS Report R41175, *International Agreements on Climate Change: Selected Legal Questions*; and CRS Report R40001, *A U.S.-Centric Chronology of the International Climate Change Negotiations*, by Jane A. Leggett. Timothy M. Lenton, Hermann Held, and Elmar Kriegler, et al., "Tipping Elements in the Earth's Climate System," Proceedings of the National Academies of Sciences, vol. 105, no. 6 (February 12, 2008), pp. 1786-1793; The Royal Society, *Geoengineering the Climate: Science, Governance, and Uncertainty,* September 2009; "Nature," *Grazing Limits Effects of Ocean Fertilization*, 2009.

However, if geoengineering technologies are deployed by the United States, another government, or a private entity, several new concerns are likely to arise related to government support for, and oversight of, geoengineering as well as the transboundary and long-term effects of geoengineering. Such was the case in the summer of 2012, when an American citizen conducted a geoengineering experiment, specifically ocean fertilization, off the west coast of Canada that some say violated two international conventions.[12]

This report is intended as a primer on the policy issues, science, and governance of geoengineering technologies. The report will first set the policy parameters under which geoengineering technologies may be considered. It will then describe selected technologies in detail and discuss their status. The third section provides a discussion of possible approaches to governmental involvement in, and oversight of, geoengineering, including a summary of domestic and international instruments and institutions that may affect geoengineering projects.

Geoengineering Governance

Geoengineering technologies aim to modify the Earth's energy balance in order to reduce temperatures and counteract anthropogenic climate change through large-scale and deliberate modifications. Implementation of some of the technologies may be controlled locally, while other technologies may require global input on implementation. Additionally, whether a technology can be controlled or not once implemented differs by technology type. Little research has been done on most geoengineering methods, and no major directed research programs are in place. Peer reviewed literature is scant, and deployment of the technology—either through controlled field tests or commercial enterprise—has been minimal.[13] Most interested observers agree that more research would be required to test the feasibility, effectiveness, cost, social and environmental impacts, and the possible unintended consequences of geoengineering before deployment; others reject exploration of the options as too risky. The uncertainties have led some policymakers to consider the need and the role for governmental oversight to guide research in the short term and to oversee potential deployment in the long term. Such governance structures, both domestic and international, could either support or constrain geoengineering activities, depending on the decisions of policymakers. As both technological development and policy considerations for geoengineering are in their early stages, several questions of governance remain in play:

- What risk factors and policy considerations enter into the debate over geoengineering activities and government oversight?

- At what point, if ever, should there be government oversight of geoengineering activities?

- If there is government oversight, what form should it take?

- If there is government oversight, who should be responsible for it?

---

[12] Martin Lukacs, "World's biggest geoengineering experiment 'violates' UN rules," *The Guardian*, October 15, 2012.

[13] Research has been minimal but not absent. In 2008, a German-Indian joint research venture on ocean fertilization produced significant debate among Parties to the London Convention and the Convention on Biological Diversity before being allowed to continue. Commercially, several companies, including Climos, Planktos, and Mantria, have investigated avenues through which to use geoengineering techniques in the carbon markets by selling emission offsets for ocean fertilization and biochar sequestration. Discussion of these and other examples can be found in Mason Inman's article, "Planning for Plan B," *Nature Reports Climate Change,* Vol. 4, January 2010.

- If there is publicly funded research and development, what should it cover and which disciplines should be engaged in it?

Risk Factors

As a new and emerging set of technologies potentially able to address climate change, geoengineering possesses many risk factors that must be taken into policy considerations. From a research perspective, the risk of geoengineering activities most often rests in the uncertainties of the new technology (i.e., the risk of failure, accident, or unintended consequences). However, many observers believe that the greater risk in geoengineering activities may lie in the social, ethical, legal, and political uncertainties associated with deployment. Given these risks, there is an argument that appropriate mechanisms for government oversight should be established before the federal government and its agencies take steps to promote geoengineering technologies and before new geoengineering projects are commenced. Yet, the uncertainty behind the technologies makes it unclear which methods, if any, may ever mature to the point of being deemed sufficiently effective, affordable, safe, and timely as to warrant potential deployment.[14] Some of the more significant risks factors associated with geoengineering are as follows:[15]

- *Technology Control Dilemma.* An analytical impasse inherent in all emerging technologies is that potential risks may be foreseen in the design phase but can only be proven and resolved through actual research, development, and demonstration. Ideally, appropriate safeguards are put in place during the early stages of conceptualization and development, but anticipating the evolution of a new technology can be difficult. By the time a technology is widely deployed, it may be impossible to build desirable oversight and risk management provisions without major disruptions to established interests. Flexibility is often required to both support investigative research and constrain potentially harmful deployment.

- *Reversibility.* Risk mitigation relies on the ability to cease a technology program and terminate its adverse effects in a short period of time. In principle, all geoengineering options could be abandoned on short notice, with either an instant cessation of direct climate effects or a small time lag after abandonment. However, the issue of reversibility applies to more than just the technologies themselves. Given the importance of internal adjustments and feedbacks in the climate system—still imperfectly understood—it is unlikely that all secondary effects from large-scale deployment would end immediately. Also, choices made regarding geoengineering methods may influence other social, economic, and technological choices regarding climate science. Advancing geoengineering options in lieu of effectively mitigating GHG emissions, for example, could result in a number of adverse effects, including ocean acidification, stresses on biodiversity, climate sensitivity shocks, and other irreversible consequences.

---

[14] See The Royal Society, *Geoengineering the Climate: Science, Governance, and Uncertainty*, 61 (2009), available for download at http://royalsociety.org/Geoengineering-the-climate.

[15] Sources: "Technology control dilemma" as outlined by the Royal Society from a definition in D. Collingridge, *The Social Control of Technology*, Francis Pinter: New York, 1980. "Reversibility" and "encapsulation" as defined by the Royal Society report, op. cit. "Commercial involvement" and "public engagement" as defined by the Royal Society report as well as broached in many of the policy articles debating the acceptability of geoengineering research and implementation.

Further, investing financially in the physical infrastructure to support geoengineering may create a strong economic resistance to reversing research and deployment activities.

- *Encapsulation.* Risk mitigation also relies on whether a technology program is modular and contained or whether it involves the release of materials into the wider environment. The issue can be framed in the context of pollution (i.e., encapsulated technologies are often viewed as more "ethical" in that they are seen as non-polluting). Several geoengineering technologies are demonstrably non-encapsulated, and their release and deployment into the wider environment may lead to technical uncertainties, impacts on non-participants, and complex policy choices. But encapsulated technologies may still have localized environmental impacts, depending on the nature, size, and location of the application. The need for regulatory action may arise as much from the indirect impacts of activities on agro-forestry, species, and habitat as from the direct impacts of released materials in atmospheric or oceanic ecosystems.

- *Commercial Involvement.* The role of private-sector engagement in the development and promotion of geoengineering may be debated. Commercial involvement, including competition, may be positive in that it mobilizes innovation and capital investment, which could lead to the development of more effective and less costly technologies at a faster rate than in the public sector. However, commercial involvement could bypass or neglect social, economic, and environmental risk assessments in favor of what one commentator refers to as "irresponsible entrepreneurial behavior."[16] Private-sector engagement would likely require some form of public subsidies or GHG emission pricing to encourage investment, as well as additional considerations including ownership models, intellectual property rights, and trade and transfer mechanisms for the dissemination of the technologies.

- *Public Engagement.* The consequences of geoengineering—including both benefits and risks discussed above—could affect people and communities across the world. Public attitudes toward geoengineering, and public engagement in the formation, development, and execution of proposed governance, could have a critical bearing on the future of the technologies. Perceptions of risks, levels of trust, transparency of actions, provisions for liabilities and compensation, and economies of investment could play a significant role in the political feasibility of geoengineering. Public acceptance may require a wider dialogue between scientists, policymakers, and the public.

Policy Considerations

Since geoengineering activities are intended to affect the climate of the planet, their consequences implicate policy considerations at both the national and international level. Accordingly, whether a country or region deems these activities, and their potential consequences, acceptable will likely

---

[16] See John Virgoe's comments in the "Uncorrected Transcript of Oral Evidence," presented before the U.K. House of Commons Science and Technology Committee on January 13, 2010. Please note that the uncorrected transcript is not yet approved as a formal record of the proceedings. Transcript can be found at http://www.parliament.uk/ parliamentary_committees/science_technology/s_t_geoengineering_inquiry.cfm.

---

depend not only on the scientific and technical underpinnings for the geoengineering technology involved, but also by a range of social, legal, and political factors that vary across countries and cultures. For example, while some may view geoengineering through the lens of religious and ethical concerns about its potential impacts, others may feel that the risk of climate inaction is too great and, therefore, that constraining geoengineering activities is as morally hazardous as promoting and engaging in geoengineering activities.

Public opinion on geoengineering is difficult to gauge at this early stage. It is likely to both evolve as more information becomes available and vary depending on the particular technology being discussed. Nevertheless, a 2009 report by the United Kingdom's Royal Society,[17] which is widely considered to be the first comprehensive analysis of geoengineering technologies, has broadly identified three categories of perspectives held within the scientific community about the deployment of geoengineering technologies:

- Geoengineering is a dangerous manipulation of Earth systems and therefore intrinsically unethical;

- Geoengineering is strictly an insurance policy against major mitigation failure; and

- Geoengineering will help buy back time lost during international mitigation negotiations.

The following table identifies and explains the scientific underpinnings for many of the perspectives on geoengineering that have been articulated to date.

---

[17] The Royal Society, *Geoengineering the Climate: Science, Governance, and Uncertainty*, 61 (2009), at http://royalsociety.org/Geoengineering-the-climate.

Table 1. Scientific Underpinnings for Different Perspectives on Geoengineering

Primary Concern	Arguments in Favor	Arguments Against
Climate uncertainty	Uncertainties in the Earth's climate may lead to catastrophic climate change and necessitate the deployment of geoengineering technologies. Adequate research, modeling, field tests, and evaluation are required to accumulate empirical evidence to prepare for, and to hedge against, a crisis.	Inadequate information about the Earth's climate system creates technical, political, social, and economic risk in geoengineering activities. Deployment of unknown technologies may lead to unintended consequences. Prudence suggests that technologies should be fully vetted for potential negative environmental or social impacts prior to deployment and a "precautionary approach" should be applied to technologies that pose a threat of serious of irreversible damage. Conceivably, some technologies, once developed, could be used adversely by hostile entities and pose threats to security.
Mitigation of GHG release	Current mitigation efforts are too slow or inadequate to achieve the emission reductions needed to reduce long-term accumulation of CO_2 and avoid dangerous changes to the climate. Geoengineering technologies may be required to augment existing mitigation strategies or to replace failing ones in order to avert a potential climate crisis.	Geoengineering activities may make permissible the continuation of business-as-usual practices and weaken conventional mitigation efforts (the "Moral Hazard" argument). In terms of climate change, this may lead to some early adopters asserting that geoengineering provides "insurance" against crisis and could embolden stakeholders to act more carelessly.
Cost of geoengineering activities	The cost of geoengineering activities could be quite small compared to the economics of mitigation or adaptation strategies. Technological innovation and entrepreneurship would only lower these costs.	It is difficult to assess the true cost of geoengineering schemes due to the uncertainties of potential side effects. Investment in market mechanisms may distort geoengineering research and deployment instead of facilitating it. Countries, corporations, and even individuals with means to pursue geoengineering may be tempted to do so out of commercial or entrepreneurial considerations that bypass or neglect risk assessments of social, economic, and environmental effects.

Primary Concern	Arguments in Favor	Arguments Against
Contingency planning for climate change	Both society and the scientific community are obliged to invest enough knowledge and resources into geoengineering that it may serve as a contingency plan in case of a climate emergency. Research in and of itself is neither good nor bad, and information about geoengineering technologies is beneficial to have "on the shelf."	Societies rarely invest adequately in contingency plans. Innovative and entrepreneurial organizations seldom mobilize themselves to put complex technologies "on the shelf." Government endorsement prematurely stamps geoengineering activities as acceptable; and given the nascent state of understanding in the science, a rush toward implementation may result in potentially dangerous proposals being mistakenly promoted and potentially useful techniques mistakenly ignored.
Governance of technologies and their deployment	Without appropriate frameworks or oversight, geoengineering technologies could be researched and deployed unilaterally by public or private actors to the detriment of other countries or populations that did not consent to the geoengineering project. In that situation, those harmed could find themselves unable both to remedy the harm and to hold the actor responsible liable for the damage suffered. Moreover, absent governance, geoengineering activities could be "spatially heterogeneous," that is, disproportionately impact particular populations and ecosystems.	If governments choose to ban or substantially restrict geoengineering, they might be constraining those actors most likely to test, assess, and deploy the technologies responsibly and, therefore leave geoengineering in the hands of the least transparent and least trustworthy actors. Moreover, too much government involvement could stifle experimentation, innovation, and entrepreneurship in technologies that could prove vital to averting excessive global warming.

Source: Congressional Research Service.

Geoengineering Technologies

A wide range of geoengineering technologies have been proposed to address climate change. Geoengineering technologies attempt to mitigate continued warming of the Earth's climate. The technologies vary in complexity from planting trees for carbon sequestration to launching mirrors into space for sunlight reflection. Most of the technologies are not yet proven and are at the theoretical or research phase. Several of the proposed technologies were recently conceived; if they prove feasible and effective, they would require large amounts of funding for full-scale deployment; and generally they lack political, scientific, and public support.

Figure 1. Geoengineering Technology Options

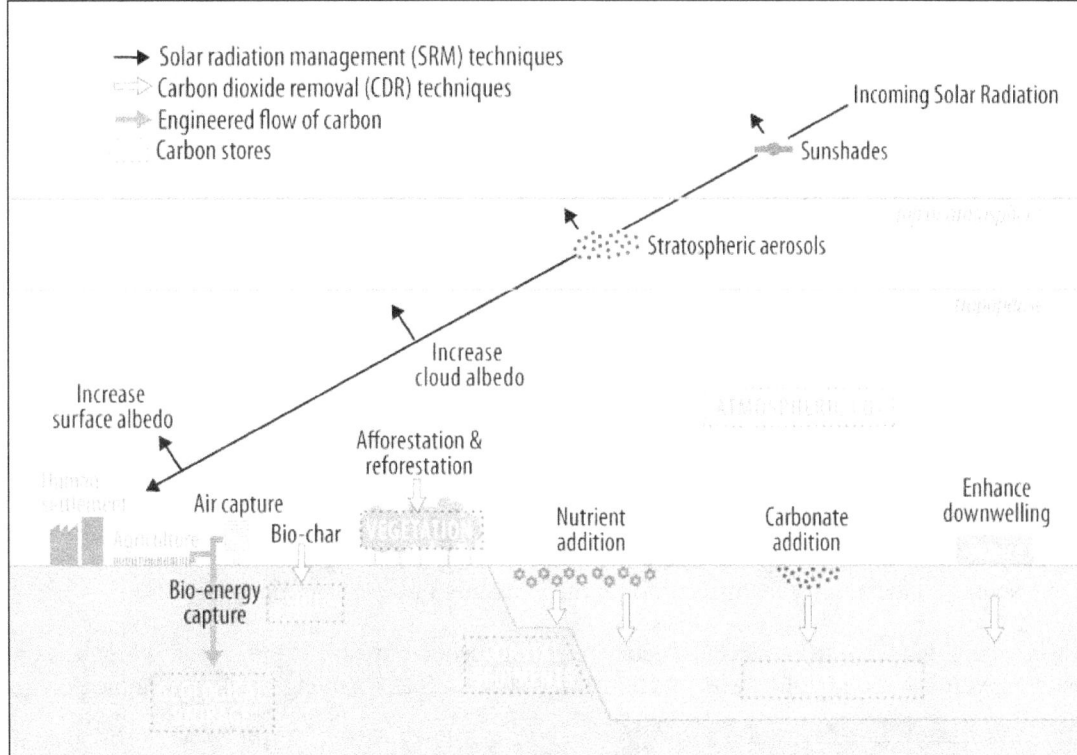

Source: T. M. Lenton and N.E. Vaughan, "The Radiative Forcing Potential of Different Climate Geoengineering Options," *Atmospheric Chemistry and Physics*, vol. 9, no. 15 (August 6, 2009). Adapted by CRS.

Note: The figure displays engineered flow of carbon as a separate geoengineering technology option. However, this report considers "engineered flow of carbon" technologies as a CDR technique.

The two main categories of geoengineering technologies are carbon dioxide removal (CDR) and solar radiation management (SRM) (see **Figure 1**). CDR methods remove CO_2 from the atmosphere. SRM methods increase the reflectivity of the Earth's atmosphere or surface, thus reducing incoming solar radiation.[18]

---

[18] House of Commons (U.K.) Science and Technology Committee, "The Regulation of Geoengineering," Fifth Report of Session 2009-10, March 18, 2010, p. 20, http://www.parliament.the-stationery-office.co.uk/pa/cm200910/cmselect/ (continued...)

Carbon Dioxide Removal

Carbon dioxide is the primary GHG that has been linked to increases in atmospheric temperature. CDR methods remove CO_2 from the atmosphere and are intended to cool the planet by reducing the absorption of heat in the atmosphere. CDR methods involve the uptake and storage of CO_2 by biological, physical, or chemical means. CO_2 may be stored (or sequestered) via afforestation,[19] ocean fertilization, weathering of certain sedimentary rocks, or combining carbon capture and storage technology with the production of biofuels, among other approaches. The duration of carbon storage differs depending on the approach. For instance, carbon storage may last from decades to centuries for ocean fertilization, or thousands of years for weathering of carbonate rock.[20]

Although CDR techniques could lower atmospheric CO_2 levels eventually leading to a decrease in temperatures, they would require considerably more time to have an impact on climate change than SRM techniques.[21] Thus, CDR techniques are not ideal methods to deploy if immediate alteration of the climate is necessary. While the impacts of CDR methods could take years to realize, many CDR methods could be governed more easily than SRM methods by existing laws. For example, carbon capture and storage from a biomass power plant could be subject to the same environmental and energy laws as carbon capture and storage from a coal-fired power plant.

This section describes some of the better-examined CDR methods: carbon capture and sequestration, ocean fertilization, afforestation, and enhanced weathering.

Carbon Capture and Sequestration

Carbon capture and sequestration (CCS) is the capture and storage of CO_2 to prevent it from being released to the atmosphere. CCS generally refers to the process of using technology to remove CO_2 released from anthropogenic sources rather than CO_2 that is captured naturally as part of the Earth's carbon cycle.[22] Because CO_2 emissions from anthropogenic sources continue to increase CO_2 levels in the atmosphere, CCS engineering technologies are being pursued for three sources: biomass, bioenergy, and fossil fuels (mainly power generation).

CCS technology tends to be labeled a geoengineering technology only if the source is biomass or bioenergy. It is not clear why the distinction of labeling CCS as a geoengineering technology

---

(...continued)

cmsctech/221/22102.htm.

[19] Afforestation is the creation of forests on land that has not recently been, or has never been, forest land.

[20] The Royal Society, *Geoengineering the Climate: Science, Governance, and Uncertainty*, 61 (2009), available for download at http://royalsociety.org/Geoengineering-the-climate.

[21] U.S. Congress, House Committee on Science and Technology, Subcommittee on Energy and Environment, *Geoengineering II: The Scientific Basis and Engineering Challenges*, statement of David Keith Canada Research Chair in Energy and Environment University of Calgary, 111th Cong., 2nd sess., February 4, 2010.

[22] Vegetation and plankton naturally capture CO_2 and sequester it in matter. A significant fraction of the CO_2 released from fossil fuel combustion is being stored in "natural" carbon sinks in the oceans and land vegetation.

CCS is a technology that can sequester large quantities of CO_2 from point sources of GHG emissions, mainly fossil fuel combustion. CCS would reduce the amount of CO_2 released into the atmosphere while allowing fossil fuel use to continue. For more information on CCS from fossil fuels, see CRS Report R42532, *Carbon Capture and Sequestration (CCS): A Primer*, by Peter Folger.

---

depends on the source from which carbon will be captured, and not its outcome, which is the reduction in the amount of CO_2 released to the atmosphere.[23] According to this view, CCS from fossil fuels may be excluded as a geoengineering technology because fossil fuels are carbon-positive, while bioenergy and biomass are viewed as carbon-neutral or carbon-negative.[24]

Additional intervention could increase the amount of CO_2 that is sequestered. Biomass that otherwise would not be used for crops or energy production could be buried in the land or deep ocean to slow the rate at which CO_2 is released to the atmosphere. Carbon can also be sequestered if biomass is pyrolyzed into biochar: a charcoal produced under high temperatures using crop residues, animal manure, or other organic material with the potential to sequester carbon in the soil for hundreds to thousands of years at an estimate.[25] Large-scale carbon capture that uses biomass would require a steady biomass supply and a place to store it.

Capturing CO_2 is relatively well understood compared to sequestering CO_2. However, CCS technology for fossil fuels has not been applied at the commercial level thus far due to lack of an economic incentive or a regulatory requirement to reduce CO_2 emissions. Carbon dioxide has been captured on a small scale for commercial applications for a few decades (e.g., for soda production, to enhance oil and gas recovery), but not on a large scale to sequester CO_2 as a long-term climate change mitigation method. Estimates of CO_2 sequestration performance are based partially on oil and gas recovery efforts which have sequestered CO_2 for approximately 30 years. It may take at least 10-15 years before CCS for fossil fuels is ready for commercial deployment.[26]

In addition to crop-based carbon capture, bioenergy generation coupled with CO_2 capture and sequestration (BECS) could sequester carbon. BECS consists of three phases: planting and growing a biomass crop such as switchgrass, harvesting the crop for biofuel production, and capturing and storing the carbon released during this process.[27] BECS is expected to use technology similar to CCS technology used for capturing CO_2 from fossil fuel combustion.[28] When biomass is used to generate electricity, the CO_2 released in the process may be sequestered in geologic formations, in the same way as it would be used in a fossil-fuel generation CCS operation.

BECS is an unproven CDR method because no commercial-scale CCS facility exists for either fossil fuels or bioenergy. Many of the lessons learned from CCS in the fossil fuels sector over the next few years should be applicable to BECS. BECS deployment could take as long or longer

---

[23] House of Commons (U.K.) Science and Technology Committee, "The Regulation of Geoengineering," Fifth Report of Session 2009-10, March 18, 2010, p. 20, http://www.parliament.the-stationery-office.co.uk/pa/cm200910/cmselect/ cmsctech/221/22102.htm.; The Royal Society, *Geoengineering the Climate: Science, Governance and Uncertainty*, September 2009.

[24] For more information on the carbon neutrality of one type of bioenergy, see CRS Report R41603, *Is Biopower Carbon Neutral?*, by Kelsi Bracmort.

[25] For more information on biochar, see CRS Report R40186, *Biochar: Examination of an Emerging Concept to Sequester Carbon*, by Kelsi Bracmort. For more information on agricultural practices that sequester carbon, see CRS Report RL33898, *Climate Change: The Role of the U.S. Agriculture Sector*, by Renée Johnson.

[26] U.S. Government Accountability Office, Coal Power Plants: Opportunities Exist for DOE to Provide Better Information on the Maturity of Key Technologies to Reduce Carbon Dioxide Emissions, GAO-10-675, June 2010, http://www.gao.gov/new.items/d10675.pdf.

[27] Peter Read and Jonathan Lermit, "Bio-energy with Carbon Storage (BECS): A Sequential Decision," *Energy*, 2005, pp. 2654-2671.

[28] Past Congresses have considered several bills to promote the development and deployment of CCS from fossil fuel sources.

than fossil-fuel CCS. Some contend that BECS could not be deployed fast enough to have a significant impact on climate change.[29]

One of the main challenges to CCS deployment is the lack of a regulatory framework to permit geologic sequestration of CO_2. An integrated structure would be necessary to deploy CCS at a large scale, whether for fossil fuels or bioenergy. This structure involves identifying who owns the sequestered CO_2, where to sequester the CO_2, defining what constitutes leakage, identifying who will be held liable if the sequestered CO_2 leaks, developing a monitoring and maintenance plan, and developing a robust pipeline infrastructure specifically for CO_2 that will be sequestered, among other things.[30] Some contend that if CCS were implemented on a large scale for both fossil fuels and bioenergy, there would be less motivation to reduce the use of fossil fuels. An increase in BECS, however, might not face the same argument. Further, BECS might be considered "carbon-negative" whereas CCS from fossil fuel combustion is at least slightly carbon-positive.[31] Additionally, there is concern that CO_2 storage from fossil fuels, and perhaps bioenergy, may lead to contamination of underground sources of drinking water. In 2010, the U.S. Environmental Protection Agency finalized a rule that sets requirements for geologic sequestration of carbon dioxide, using the authority granted the agency in the 1974 Safe Drinking Water Act.[32]

Ocean Fertilization

Ocean fertilization is the addition of nutrients such as iron to the ocean to expedite carbon sequestration from phytoplankton.[33] Phytoplankton photosynthesize CO_2, retaining the carbon in their cells, which then is sequestered as carbon in the deep ocean when they die and settle through the waters. Studies suggest that a ton of iron added to certain parts of the ocean could remove 30,000 to 110,000 tons of carbon from the air.[34] Ocean fertilization is estimated to cost approximately $30 to $300 per ton of carbon sequestered.[35]

---

[29] Christian Azar, Kristian Lindgren, and Eric Larson, "Carbon Capture and Storage from Fossil Fuels and Biomass—Costs and Potential Role in Stabilizing the Atmosphere," *Climatic Change*, vol. 74, no. 1-3 (2006).

[30] For more information on pipeline construction challenges for CCS, see CRS Report RL33971, *Carbon Dioxide (CO2) Pipelines for Carbon Sequestration: Emerging Policy Issues*, by Paul W. Parfomak, Peter Folger, and Adam Vann.

[31] Carbon-positive fuels are drawn from fossil fuel deposits and are burned, releasing CO2 into the atmosphere. Carbon-neutral fuels absorb CO2 as they grow and release the same carbon back into the atmosphere when burnt. Carbon-negative fuels absorb CO2 as they grow and release less than this amount into the atmosphere when used as fuel, either through directing part of the biomass as biochar back into the soil or through CCS.

[32] U.S. Environmental Protection Agency, "EPA Finalizes Rules to Foster Safe Carbon Storage Technology Actions," press release, November 22, 2010, http://yosemite.epa.gov/opa/admpress.nsf/0/ 2300005FBC11568D852577E3006058BD. The Safe Drinking Water Act gives the U.S. Environmental Protection Agency (EPA) the authority to regulate underground injections of numerous substances. For more information on the Safe Drinking Water Act and the Underground Injection Control Program, see CRS Report RL34201, *Safe Drinking Water Act (SDWA): Selected Regulatory and Legislative Issues*, by Mary Tiemann.

[33] Iron is the primary nutrient discussed in literature for ocean fertilization, although other nutrients (e.g., nitrogen) may be used. References to ocean fertilization in this report refer to iron as the fertilizing nutrient unless otherwise noted.

[34] "Fertilizing the Ocean with Iron: Should We Add Iron to the Sea to Help Reduce Greenhouse Gases in the Air," *Oceanus*, November 13, 2007.

[35] Philip W. Boyd, "Implications of large-scale iron fertilization of the oceans," *Marine Ecology Progress Series*, vol. 364 (July 29, 2008), pp. 216-217.

The ecological, economic, and climatological impacts of ocean fertilization, in both the short term and the long term, are uncertain.[36] Some suggest that ocean fertilization may enhance fish stocks and augment production of dimethylsulfide, a chemical that may cool the atmosphere, but has undesirable characteristics at high concentrations.[37] Others are concerned that ocean fertilization will lead to ocean acidification, additional emissions of potent greenhouse gases, and reduction of oxygen to levels not habitable by certain species. Critics also argue that ocean fertilization is not an effective way to combat climate change because the technique requires widespread long-term implementation on a continual basis.[38]

Studies have yet to demonstrate that ocean fertilization will work as a long-term carbon sequestration strategy. Further research is likely needed to answer numerous questions: Will phytoplankton increase in sufficient numbers to sequester significant amounts of CO_2? How long will the carbon stay sequestered? What disruptions will occur to marine ecosystems? Currently, there are no analogues to compare what may occur if ocean fertilization is deployed on a large scale.

Some envision CO_2 sequestered via ocean fertilization as a potential carbon credit to be sold as a carbon offset or traded within an environmental market.[39] There appear to be no legal frameworks that endorse or reject ocean fertilization for the purpose of acquiring carbon credits. Thus, for the time being, any carbon credits garnered for ocean fertilization would have to be used in a voluntary carbon market.[40]

Afforestation

Afforestation involves planting tree seedlings on sites that have been without trees for several years, generally a decade or more.[41] The primary climate change benefit of afforestation discussed in scientific and policy literature is carbon sequestration. It is regarded as a prime carbon sequestration strategy because forest communities can store about 10 times more carbon in their vegetation than non-forest communities and for longer time periods (decades to hundreds of years). Other benefits include erosion control, recreational value, wildlife habitat, and production of forest goods. On a large scale, afforestation can modify local climates by increasing humidity, altering cloud and precipitation patterns, and reducing wind speeds. Challenges associated with afforestation include measurement and reporting of carbon storage, landowners'

---

[36] R. Sagarin et al., "Iron Fertilization in the Ocean for Climate Mitigation: Legal, Economic, and Environmental Challenges," Duke University, Nicholas Institute for Environmental Policy Solutions, 2007.

[37] "Fertilizing the Ocean with Iron: Should We Add Iron to the Sea to Help Reduce Greenhouse Gases in the Air," *Oceanus*, November 13, 2007. Dimethyl sulfide has an unpleasant odor at low concentrations, and is flammable and an eye irritant at high concentrations.

[38] Aaron Strong, Sallie Chisholm, and Charles Miller et al., "Ocean Fertilization: Time to Move On," *Nature*, vol. 461 (September 17, 2009), pp. 347-348; and Aaron L. Strong, John J. Cullen, and Sallie W. Chisolm, "Ocean Fertilization: Science, Policy, and Governance," *Oceanography*, vol. 22, no. 3 (2009).

[39] Ocean fertilization is not recognized as a creditable offset under the Kyoto Protocol for the first commitment period effective through 2012. For more information on carbon offsets, see CRS Report RL34241, *Voluntary Carbon Offsets: Overview and Assessment*, by Jonathan L. Ramseur, and CRS Report RL34436, *The Role of Offsets in a Greenhouse Gas Emissions Cap-and-Trade Program: Potential Benefits and Concerns*, by Jonathan L. Ramseur.

[40] "Dumping Iron and Trading Carbon: Profits, Pollution, and Politics All Will Play Roles in Ocean Iron Fertilization," *Oceanus*, January 10, 2008.

[41] For more information on afforestation, see CRS Report RL34560, *Forest Carbon Markets: Potential and Drawbacks*, by Jonathan L. Ramseur.

---

reluctance to grow trees on private land for extensive time periods, and a reduction in runoff that may impact the ecology of the afforested area, among other issues.[42] Potential drawbacks to wide-scale implementation of afforestation include unexpected releases of CO_2 from newly forested lands due to acts of nature (e.g., fire, drought), future changes in land management (e.g., harvesting) that could result in release of the carbon, the potentially significant cost for afforestation, and possible effects on crop production and agricultural commodity prices if significant croplands are afforested.

The planting of trees is well known, and well practiced; afforestation is an accepted project activity under the Clean Development Mechanism (CDM) of the Kyoto Protocol.[43] The amount and rate at which CO_2 is sequestered depends on the tree species, climate, soil type, management, and other site-specific features. The estimated sequestration potential ranges from 2.2 to 9.5 metric tons of CO_2 per acre per year.[44] It may take at least 20 years to reap the carbon sequestration benefit depending on the growth rate of the trees. Carbon accumulation in the early years of tree growth is slow and increases during the strong growth period; there is controversy whether carbon accumulation continues or peaks when net additional wood growth is minimal.[45]

Most afforestation projects occur on marginal croplands. Certain models estimate that a total of 60 million to 65 million acres of U.S. agricultural land could be converted to woodlands by 2050, including 35 million to 50 million acres of cropland.[46] Changes in climate may impact which forestry species are planted at afforestation sites. Therefore, forestry species planted at afforestation projects may not be similar to species found at the site in the past. The cost of an afforestation project can range from approximately $65 to $200 per acre due in part to the previous land use of the site and the terrain.[47]

Enhanced Weathering

Carbon dioxide is naturally removed from the atmosphere slowly through weathering (or disintegration) of silicate and carbonate rocks. Expediting the weathering process—enhanced weathering—could remove large amounts of CO_2 from the atmosphere.[48] The disintegrated materials containing CO_2 removed from an enhanced weathering project could be stored in the deep ocean or in soils.

---

[42] Kathleen A. Farley, Esteban G. Jobbagy, and Robert B. Jackson, "Effects of Afforestation on Water Yield: A Global Synthesis with Implications for Policy," *Global Change Biology*, vol. 11 (2005), pp. 1565-1576. For more information on the impacts of afforestation, see CRS Report R41144, *Deforestation and Climate Change*, by Pervaze A. Sheikh.

[43] The Clean Development Mechanism (CDM) allows a country with an emission-reduction or emission-limitation commitment under the Kyoto Protocol to implement an emission-reduction project in developing countries. Such projects can earn saleable certified emission reduction (CER) credits, which can be counted towards meeting Kyoto targets. The mechanism stimulates sustainable development and emission reductions, while giving industrialized countries some flexibility in how they meet their emission reduction or limitation targets.

[44] U.S. Environmental Protection Agency, Office of Atmospheric Programs, *Greenhouse Gas Mitigation Potential in U.S. forestry and Agriculture*, EPA 430-R-05-006, Washington, DC, November 2005, Table 2-1.

[45] For more information, see CRS Report R40562, *U.S. Tree Planting for Carbon Sequestration*, by Ross W. Gorte.

[46] For more information on afforestation as a carbon offset practice, see CRS Report R41086, *Potential Implications of a Carbon Offset Program to Farmers and Landowners*, by Renée Johnson et al.

[47] Lucas S. Bair and Ralph J. Alig, *Regional Cost Information for Private Timberland Conversion and Management*, U.S. Department of Agriculture., PNW-GTR-684, September 2006, http://www.fs.fed.us/pnw/pubs/pnw-gtr684.pdf.

[48] "Silicate Rocks React with CO_2 to Form Carbonates Thus Consuming CO_2." The Royal Society, *Geoengineering the Climate: Science, Governance, and Uncertainty*, September 2009.

A paucity of literature exists about how to conduct an enhanced weathering project or its environmental implications. One proposed method is to spread crushed olivine, a type of silicate rock, on agricultural and forested lands to sequester CO_2 and improve soil quality.[49] This technique would require large amounts of rocks to be mined, ground, and transported. The lifecycle carbon benefit has not been calculated. Significant amounts of additional resources, such as energy and water, may be required to conduct an enhanced weathering project.

Further research would be needed to assess the potential benefits and drawbacks of this technology. Barriers to enhanced weathering include its scale, cost, energy requirements, and potential environmental consequences.[50] Decisions would need to be made on which landscape to alter, where to dispose of the disintegrated material, and who pays for the project. There may be long-term adverse impacts on air quality, water quality, and aquatic life.

Solar Radiation Management

Solar radiation management methods work to reduce or divert the amount of incoming solar radiation by making the Earth more reflective (i.e., enhancing albedo) and do not have any effect on GHG emission rates.[51] SRM methods involve modifying albedo via land-based methods such as desert reflectors, cloud-based methods such as cloud whitening, stratosphere-based methods such as aerosol injection, and spaced-based methods such as shields. The effectiveness of an SRM method depends on its geographical location, the altitude at which it is applied (surface, atmosphere, space), and the radiative properties of the atmosphere and surface.

If proven effective and desirable, SRM methods could be deployed faster than CDR methods should the need arise to cool the planet quickly. SRM methods have been described, theoretically, as cheap, fast, and imperfect.[52] However, these methods have not been proven on any scale. Some argue the U.S. government should create a research or oversight program potentially with international cooperation that examines SRM technologies prior to a potentially hasty deployment by an individual or country, which could result in an array of unanticipated consequences.[53] Research could improve understanding of the feasibility of different SRM approaches, their opportunities and limitations, and their potential role in climate change mitigation.[54] Other commentators favor constraints on SRM research, given the significant environmental risks posed by these techniques:

- System failure. If an SRM technique breaks down or is shut down, the climate may warm very quickly, possibly leaving little time for humans and nature to adapt.

---

[49] R. D. Schuiling and P. Krijgsman, "Enhanced Weathering: An Effective and Cheap Tool to Sequester CO_2," *Climatic Change*, vol. 74 (2006), pp. 349-354.

[50] The Royal Society, *Geoengineering the Climate: Science, Governance, and Uncertainty*, 61 (2009), available for download at http://royalsociety.org/Geoengineering-the-climate.

[51] Ibid.

[52] David W. Keith, Edward Parson, and M. Granger Morgan, "Research on Global Sun Block Needed Now," *Nature*, vol. 463 (January 28, 2010), pp. 426-427.

[53] Ibid.

[54] Lee Lane, Ken Caldeira, and Robert Chatfield, et al., *Workshop Report on Managing Solar Radiation*, National Aeronautics and Space Administration , NASA/CP-2007-214558, Hanover, MD, April 2007, http://event.arc.nasa.gov/main/home/reports/SolarRadiationCP.pdf.

- Changes in regional and seasonal climates. SRM techniques may alter precipitation patterns, which could have consequences for ecosystems and affected societies.

- Ozone depletion. Under certain circumstances, use of SRM techniques such as sulfate aerosol injection may lead to ozone depletion which would allow harmful UVB rays to reach the Earth.

- Preservation of non-CO_2 greenhouse gases. SRM techniques applied in the stratosphere or space lessen the amount of ultraviolet radiation striking the Earth's atmosphere, which is likely to extend the atmospheric lifetime of non-CO_2 greenhouse gases that are more potent than CO_2.

- Diversion from more permanent solutions. If societies conclude that SRM techniques can provide quick relief, they may invest less in developing and deploying more permanent GHG emission reduction solutions.

- "Unknown unknowns." The history of the Earth's climate demonstrates that small changes may result in abrupt changes, raising concerns about unknown effects of large-scale geoengineering.[55]

The following section explores some of the more widely discussed SRM techniques: enhanced albedo, aerosol injection, and space-based reflectors.

Enhanced Albedo (Surface and Cloud)

One suggested method to modify the temperature of the planet is to increase the reflectivity, or albedo, of certain surfaces. Increasing surface reflectivity directs more solar radiation back toward space thus limiting temperature increases. Surface types, application areas, and costs for enhanced albedo are all under investigation.

One of the most widely discussed targets for enhancing surface albedo is urban areas. Applying enhanced albedo methods in urban areas such as painting roofs and paved areas white on a global basis is estimated to cost several billion dollars for materials and labor, but could save money on energy costs.[56] For example, the U.S. Department of Energy (DOE) National Nuclear Security Administration (NNSA) has reduced building heating and cooling costs by an average of 70 percent annually on reroofed areas partly due to installing cool roofs.[57] Some drawbacks to increased reflectivity of roofs and paved areas include uncomfortable glare, concern for the aesthetic appeal of the roof or paved area depending on its location, the loss of reflectivity benefits if the roof is poorly maintained, and increased energy costs in colder climates due to

---

[55] First four constraints attributed to Lane et. al (2007). For more information on non-CO_2 greenhouse gases, see CRS Report R40813, *Methane Capture: Options for Greenhouse Gas Emission Reduction*, by Kelsi Bracmort et al., and CRS Report R40874, *Nitrous Oxide from Agricultural Sources: Potential Role in Greenhouse Gas Emission Reduction and Ozone Recovery*, by Kelsi Bracmort.

[56] Hashem Akbari, Surabi Menon, and Arthur Rosenfeld, "Global Cooling: Increasing World-Wide Urban Albedos to Offset CO_2," *Climatic Change*, vol. 94, no. 3-4 (2009), pp. 275-286.

[57] Department of Energy, "Secretary Chu Announces Steps to Implement Cool Roofs at DOE and Across the Federal Government," press release, July 19, 2010, http://www.energy.gov/news/9225.htm. Cool roofs are roofs that are designed to maintain a lower roof temperature than traditional roofs while the sun is shining.

---

reduced beneficial winter time heat gains.[58] Additionally, if enhanced surface albedo for paved areas is pursued aggressively, there may be a decline in the use of asphalt—a petroleum residue.

Additional techniques are being considered for enhancing surface albedo. One proposal is to modify plants through genetic engineering to augment albedo with relatively low implementation costs.[59] Some maintain it will take at least a decade for enhanced albedo plant varieties to be available commercially.[60] A second proposal is covering oceans with reflective surfaces to enhance albedo. There are concerns about where an enhanced albedo project would take place in the ocean and what impact it would have on aquatic life.

Cloud whitening is another proposed method for enhancing albedo. Cloud whitening is the dispersion of cloud-condensation nuclei (e.g., small particles of sea salt) in clouds in desired areas on a continual basis (see **Figure 2**). Aircraft, ships, or unmanned, radio-controlled seacraft could disperse the nuclei.[61] Satellites have been proposed as a way to measure cloud albedo and determine the amount of cooling needed. Spraying for cloud whitening could be halted quickly if unexpected consequences arose with cloud properties expected to return to normal within a few days.[62]

The long-term implications of deploying cloud whitening are not yet fully understood. Depending on the scale of the project, marine ecosystems could be disturbed. Further research is needed for spray generator development, and to assess potential impacts on ocean currents and precipitation patterns. Moreover, the amount of cooling that could take place and at which locations requires greater study. One study identified the west coast of North America, among other locations, as an area where cloud albedo might be effectively enhanced.[63]

---

[58] Bryan Urban and Kurt Roth, *Guidelines for Selecting Cool Roofs*, U.S. Department of Energy, July 2010, http://www1.eere.energy.gov/femp/pdfs/coolroofguide.pdf.

[59] Andy Ridgwell, Joy S. Singarayer, and Alistair M. Hetherington, et al., "Tackling Regional Climate Change by Leaf Albedo Bio-geoengineering," *Current Biology*, vol. 19 (January 27, 2009). The authors propose to genetically modify plant leaf or canopy structure to achieve greater temperature reductions.

[60] Joy S. Singarayer, Andy Ridgwell, and Peter Irvine, "Assessing the Benefits of Crop Albedo Bio-geoengineering," *Environmental Research Letters*, vol. 4 (2009).

[61] The Royal Society, *Geoengineering the Climate: Science, Governance, and Uncertainty*, 61 (2009), available for download at http://royalsociety.org/Geoengineering-the-climate.

[62] John Latham, Philip Rasch, and Chih-Chieh Chen, et al., "Global Temperature Stabilization via Controlled Albedo Enhancement of Low-level Maritime Clouds," *Philosophical Transaction of the Royal Society*, vol. 366 (August 29, 2008), pp. 3969-3987.

[63] Ibid.

Figure 2. Cloud Whitening Schematic

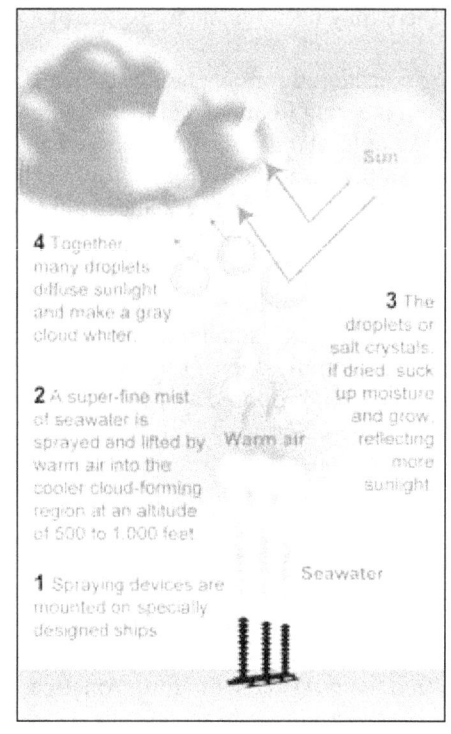

Source: Oren Dorell, "Can Whiter Clouds Reduce Global Warming?," *USA Today*, June 11, 2010. Adapted by CRS.

Aerosol Injection

Aerosol injection is the dispersal of aerosols, such as hydrogen sulfide (H_2S) or sulfur dioxide (SO_2), into the stratosphere to direct solar radiation back toward space or absorb heat, thus cooling the Earth.[64] Military aircraft, artillery shells, or stratospheric balloons could be employed to inject the aerosols. The annual cost for sulfur particle injection using airplanes is calculated to be several billion dollars, depending on the amount, location, and type of sulfur particle injected into the stratosphere.[65] However, there has not been any testing to determine whether the theoretical predictions will match reality.

Aerosol injection seeks to imitate large volcanic eruptions. Indeed, many studies have based aerosol injection simulations on data gathered and analyzed from the Mount Pinatubo volcanic eruption in the Philippines in 1991, which led to a reduction in global temperatures, though not distributed evenly across regions.[66] Sulfur releases from volcanic eruptions are random, with

---

[64] Sulfate particles are the primary injectant discussed in the literature on aerosol injection, although other particles may some day be studied and recommended for use.

[65] Alan Robock, Allison Marquardt, and Ben Kravitz et al., "Benefits, Risks, and Costs of Stratospheric Geoengineering," *Geophysical Research Letters*, vol. 36 (October 2, 2009).

[66] Ingo Kirchner, Georgiy L. Stenchikov, and Hans-F. Graf et al., "Climate Model Simulation of Winter Warming and Summer Cooling Following the 1991 Mount Pinatubo Volcanic Eruption," *Journal of Geophysical Research*, vol. 104, no. D16 (August 27, 1999), pp. 19039-19055; and Kevin E. Trenberth and Aiguo Dai, "Effects of Mount Pinatubo Volcanic Eruption on the Hydrological Cycle as an Analog of Geoengineering," *Geophysical Research Letters*, vol. 34 (continued...)

cooling impacts that have lasted no more than a few years. Aerosol injection would probably have to occur several times over decades or centuries to offset radiative forcing caused by greenhouse gases due to the short effectiveness time frame of aerosol injection.[67]

The benefits and risks of aerosol injection would not be evenly distributed around the globe. A potential benefit, in addition to cooling of the planet, could be reduced or reversed sea and land ice melting (as long as the aerosols don't settle on and darken snow and ice).[68] Some risks could be drought in Africa and Asia leading to a loss in agricultural productivity, the GHG impact that would accumulate from transporting the aerosol to the site of injection, stratospheric ozone depletion, weakening of sunlight for solar power, a less blue sky, and obstruction of Earth-based optical astronomy.[69]

Space-Based Reflectors

Space-based reflectors—a theoretical geoengineering technology proposal—would be shields positioned in space to reduce the amount of incoming solar radiation. The effectiveness of the shield would vary based on its design, material, location, quantity, and maintenance. The types of shield materials that have been suggested are lunar glass, aluminum thread netting, metallic reflecting disks, and refracting disks.[70] Proposed shield locations include the low Earth orbit and Lagrange point 1 (L1).[71]

Many aspects of using space-based reflectors require additional study. In particular, further research is needed to assess shield costs; appropriate steps for implementation, including transportation to the desired location; maintenance needs; shield disposal; and ecological impacts. Several questions have yet to be answered: Would the space-based reflectors be deployed to alter the climate at a global or regional level? Is the science behind reflector deployment mature enough to provide guidance on where shield protection would be most needed? It may take several decades to construct and deploy a shield. Should the shield fail or be removed, warmer temperatures would ensue rapidly if CO_2 emission rates continued to rise. One study suggests that launching a shield to fully reverse global warming may cost a few trillion dollars, implemented over a 25-year time frame.[72]

(...continued)

(August 1, 2007).

[67] The Royal Society, *Geoengineering the Climate: Science, Governance, and Uncertainty*, 61 (2009), available for download at http://royalsociety.org/Geoengineering-the-climate.

[68] Alan Robock, Allison Marquardt, and Ben Kravitz, et al., "Benefits, Risks, and Costs of Stratospheric Geoengineering," *Geophysical Research Letters*, vol. 36 (October 2, 2009).

[69] Ibid.

[70] The Royal Society, *supra* note 14.

[71] Lagrange points are imaginary points in space at which objects sent from Earth will stay put. Lagrange point 1 (L1) is located about four times farther from Earth than the moon.

[72] Roger Angel, "Feasibility of Cooling the Earth with a Cloud of Small Spacecraft Near the Inner Lagrange Point (L1)," *Proceedings of the National Academy of Sciences of the United States of America*, vol. 103, no. 46 (November 14, 2006), pp. 17184-17189.

The Debate over the Methods of Oversight

Geoengineering is an emerging policy area. The decision of policymakers to either pursue or constrain geoengineering research and/or deployment activities may be based on a wide assortment of factors, including social, legal, and political factors as well as scientific and technical ones—not the least of which may be progress on other climate-related policies. The debate over the point at which governments may oversee geoengineering activities as well as the form which potential oversight may take is presented in the following section.

The Status Quo

One possible governmental response to geoengineering activities is to continue the status quo. While the status quo varies depending on the country and technology being discussed, it can broadly be described as the provision of modest funding for geoengineering research, the limited regulation of particular geoengineering activities, and a lack of a comprehensive system of oversight or technology promotion. Advocates of maintaining this status quo tend to see private industry and commercial development as the best avenue through which to pursue geoengineering research and entrepreneurship. They may contend that the commercial sector is structured to promote innovation, and that current and proposed regulation and/or legislation for greenhouse gas emission reductions (or lack thereof) is fully adequate to push the market toward the development of climate engineering, if necessary. Advocates of the status quo may stress that currently funded research in marine and atmospheric science, carbon capture and sequestration, and adaptation strategies can lay sufficient groundwork for technological transformation, and that significant investment in a contingency strategy or an "insurance policy" is economically misplaced. Greater government intervention in geoengineering research and development may stifle fact gathering and agenda setting, incorrectly choose winners and losers, and unnecessarily circumscribe a science still in its infancy. Additionally, those who want to avoid governmental support of geoengineering may consider new methods of support an inappropriate stamp of acceptability on a technology some consider deleterious. As such, further support may be unwarranted because public opinion and civic engagement may have already soured to either the technological uncertainty of the science or the cost considerations of research and regulation.

If governments opt to address geoengineering activities without engaging in new law or treaty making, they would essentially endorse the status quo. This, however, appears unlikely since policymakers have begun to discuss a greater need for governance. Nevertheless, the arguments for governmental abstention will play a role as governments begin to determine the appropriate thresholds for, types, and extent of government interventions in geoengineering.

Threshold for Oversight

If policymakers decide to address geoengineering more aggressively or comprehensively, one possible policy proposal will entail creating a system for government oversight of both research and deployment of geoengineering technologies. The question of *when* governments should intervene to monitor or regulate geoengineering activities will be key for determining an appropriate system of oversight. In other words, policymakers must identify the particular harms or effects that they may wish to monitor or regulate and then determine the *de minimis* level of geoengineering activity that is likely to cause them. Observers have suggested some criteria for determining the point at which geoengineering activities should become subject to a larger system of oversight or regulation.[73] These criteria include:

- the extent to which the impacts of geoengineering *are transboundary or international in scope*;

- the extent to which the impacts of geoengineering include the introduction of *hazardous material into the environment*;

- the extent to which the impacts of geoengineering *directly intervene in the balance of ecosystems*; and

- the potential perturbation, reversibility, and duration of the geoengineering activity under discussion.

Ultimately, this assessment may require substantial scientific, as well as political, insight. Accordingly, some observers have suggested that governments set up an international panel process, similar to the one used by the National Academy of Science/National Research Council, under which scientific experts would testify and debate the merits of particular issues and approaches with the aim of providing insight to policymakers. However, this suggestion has been criticized for excluding social scientists, ethics analysts, the greater public, and other interested parties. An alternate suggestion is that governments should evaluate research and deployment activities in a manner similar to the one prescribed by the National Environmental Policy Act (NEPA), under which federal agencies prepare environmental impact statements for actions that significantly affect the quality of the human environment.[74]

Methods for Oversight

Beyond the appropriate level of public oversight for geoengineering technologies, questions concerning the potential forms of oversight may arise. It is possible that different policies may be required in order to address different concerns about the technologies. These concerns include:

- the technology is new and unproven, with ongoing and transformative scientific and technical evidence;

- the impacts of geoengineering activities are uncertain in scope, timing, and intensity;

- the range of stakeholders potentially affected by geoengineering activities is broad, including most nations, subnational groups, nongovernmental organizations, corporations, and civil societies;

- the number of actors potentially employing geoengineering activities may be small in comparison to the number of those affected;

- the global impacts of geoengineering activities—both its benefits and risks—may be unevenly distributed across stakeholders; and

---

[73] Royal Society memorandum to the U.K. House of Commons, op.cit.

[74] *Geoengineering III: Domestic and International Research Governance Hearing Before H. Comm. on Science and Technology*, 111th Cong. (2010) (written testimony of Dr. Jane C. S. Long, Associate Director at Large, Lawrence Livermore National Laboratory), available at http://democrats.science.house.gov/Media/file/Commdocs/hearings/2010/Full/18mar/Long_Testimony.pdf.

- the costs of implementing geoengineering activities may be small compared to the economics of their full global impact.

As such, different technologies, different stages of the research and deployment cycle, and different environments for research and deployment activities may require different methods for oversight.

Different technologies may require different methods for oversight. To the extent that CDR technologies are similar to known and existing ones, their development and implementation may be adequately governed at the domestic level by existing U.S. laws. Air capture technologies are similar to those of carbon capture and sequestration for power generation. Biochar and biomass sequestration face similar life cycle analyses and regulatory issues to biofuels. Ecosystem impacts of enhanced terrestrial weathering would likely be contained within national boundaries. Enhanced weathering in oceans and ocean fertilization techniques, however, are CDR technologies that may not be currently governed by U.S. law and if deployed, could cause non-trivial effects across national boundaries. Similarly, sequestration of CO_2 geologically or in the oceans may not be analogous to regulation of other underground injection or well management.[75] In addition, the scope, dispersions, and interventions of most SRM technologies are very likely to cause significant effects across national boundaries. While land surface albedo modification could potentially be managed under national regulatory frameworks, other technologies may trigger transboundary issues. While some existing treaties address atmosphere and space, their enforcement has rarely been tested.

Different stages of the research and deployment cycle may require different methods for oversight. Geoengineering development involves several stages. Regulatory frameworks must be flexible enough to cover the full cycle (from assessment through research, modeling, laboratory trials, field trials, implementation, monitoring, and evaluation). Many scientists stress a difference in regulatory needs between geoengineering research activities and full-scale deployment. But while some contend that the early stages of investigation may require little oversight, transboundary environmental impacts could grow quickly during technological development, and negative social and economic consequences may be felt as early as small-scale field trials.

Different environments for potential research or deployment may require different methods for oversight. To the extent that geoengineering technologies are governed by existing laws and treaties, they tend to fall within the provisions of legal instruments designed to regulate the uses of particular environments (e.g., space, atmosphere, ocean, land). Whereas the uses of some of these environments, such as land and the local atmosphere, are amenable to regulation via domestic laws, the uses of others, such as the high seas, transboundary and upper atmosphere, or outer space, cannot be effectively regulated without international cooperation.

Moratoriums

A final consideration for potential government oversight may be the imposition of a moratorium on geoengineering research, deployment, or both for some technologies or practices. In general,

---

[75] CRS Report RL33971, *Carbon Dioxide (CO2) Pipelines for Carbon Sequestration: Emerging Policy Issues*, by Paul W. Parfomak, Peter Folger, and Adam Vann. See also The Royal Society, *Geoengineering the Climate: Science, Governance, and Uncertainty*, 61 (2009), at http://royalsociety.org/Geoengineering-the-climate, at 51 (suggesting that different kinds of regulations will be appropriate for different methods of geoengineering).

advocates of a moratorium state that (1) the underlying science is too uncertain and too risky and should be avoided as a precaution against adverse effects to the ecosystem; (2) the potential impacts are too uneven, or disproportionately weighted toward vulnerable groups, societies, or developing countries; (3) geoengineering research would distract from the global goal of mitigation, adaptation, and emission reductions (the "moral hazard" argument); (4) geoengineering could be used by governments and industry as a "time buying" strategy to delay mitigation commitments; and (5) geoengineering techniques have the potential of being co-opted by commercial or unethical interests.

Others suggest that a moratorium on geoengineering technologies would be ill-advised. From their perspective, a moratorium would (1) inhibit research, some of which has been ongoing for decades in the context of marine ecology, oceanographic studies, and atmospheric meteorology; (2) make it difficult to accumulate the information needed to make informed judgments about the feasibility and the acceptability of the proposed technology; and (3) likely deter only those countries, corporations, and individuals who are most likely to develop the technology in a responsible fashion, thus failing to discourage potentially dangerous experimentation by less responsible parties.

The Debate over Oversight and Governmental Involvement

At present, only a few of the possible geoengineering activities appear to be under the jurisdiction of domestic laws or international treaties, and it is largely unclear how those legal instruments would encourage or constrain these activities in practice. While some continue to debate the merits of government oversight of geoengineering activities, others have suggested that new legal frameworks should be developed to support coordinated and collaborative research, develop normative standards for enforcement, and/or prevent or minimize the risks in development and deployment. Moreover, whether frameworks are best implemented at a local, national, or international level (perhaps different for different technologies) is uncertain, and debate remains over what agencies or organizations should be tasked with oversight.

The following section summarizes three different approaches to government oversight of geoengineering activities: methods of sub-national oversight, methods of national oversight, and methods of international oversight. In doing so, it also summarizes existing laws and treaties that may affect geoengineering research and deployment activities.

State Policies Addressing Geoengineering

In the United States, one possible method for the governance of geoengineering activities is to let states develop their own policies. A component of U.S. federalism is the potential for states to act as laboratories for regulatory innovation and experimentation. However, the potential effects of geoengineering across state and regional boundaries may necessitate the development of a more comprehensive federal policy. In addition, the likely diversity and plurality of state geoengineering policies may make it difficult for private actors, be they scientific researchers or corporations, who often find it easier to operate under a single uniform set of laws rather than under a multitude of different ones. An examination of current state involvement in geoengineering policy is beyond the scope of this report.

National Policies Addressing Geoengineering

Efforts by the U.S. government to develop policies addressing the use of emerging technologies are well documented.[76] These efforts may indicate best practices for developing effective policies to address the deployment of geoengineering technologies, which, like prior emerging technologies, are associated with uncertainty and a variety of social, ethical, and environmental risks.[77] From a research standpoint, emerging high-risk technologies (such as geoengineering) often struggle to obtain private sector financing and/or research support during the initial phases of investigation. Reasons for the lack of private sector involvement in geoengineering may include (1) aversion to investing in long-term technical uncertainty; (2) lack of a price mechanism on carbon to incentivize deployment of the technologies; (3) uncertainty over the commercial or private sector use of the technologies beyond large-scale government implementation; and (4) a desire not to engage until certain social, economic, and environmental risks are addressed. Consequently, emerging technologies (such as geoengineering) may require some measure of initial public subsidy to help spur development. Some such subsidies already exist at the federal level in the United States for some technologies (e.g., carbon sequestration, renewable energy). Conversely, from a regulatory standpoint, emerging technologies (such as geoengineering) carry the risk of hazard and unintended consequences. Due to the uncertainties for public health, safety, and welfare, geoengineering activities may require constraints, prohibitions, or bans comparable to the regulatory controls placed on hazardous materials or waste streams.

Current U.S. Policies Addressing Geoengineering

While no federal law has been enacted with the express purpose of covering geoengineering activities, some legal instruments may currently apply to domestic geoengineering practices and their impacts, depending on the type, location, and sponsor of the activity. The federal government could expand these existing laws to specifically address geoengineering activities or develop new laws. In addition, administrative agencies could interpret their statutory authority to authorize new rules explicitly addressing particular geoengineering activities. Among the geoengineering activities that may already be affected by existing federal laws:

- *Subterranean carbon dioxide sequestration*, which may implicate provisions of the Safe Drinking Water Act, 42 U.S.C. 300f *et. seq.*, and the Clean Air Act, 42 U.S.C. § 7401 *et. seq.* In July 2008, the U.S. Environmental Protection Agency (EPA) relied on its authority under the Safe Water Drinking Act to issue a draft rule that would regulate CO_2 injection for the purposes of geological sequestration.[78] More recently, the EPA relied on its authority under § 307(d) and

---

[76] For example, the fields of nuclear science, molecular biology, and nanotechnology represent areas in which the U.S. government has developed policies to address new and potentially risky technologies. For analysis of the science and governance behind these fields, see CRS Report RL33558, *Nuclear Energy Policy*, by Mark Holt; CRS Report RL34376, *Genetic Exceptionalism: Genetic Information and Public Policy*, by Amanda K. Sarata; and CRS Report RL34511, *Nanotechnology: A Policy Primer*, by John F. Sargent Jr. et al.

[77] *See generally* Lynn Frewer & Brian Salter, *Public Attitudes, Scientific Advice, and the Politics of Regulatory Polices: The Case of BSE*, 29 *Sci. & Pub. Pol'y* 137 (2002) (identifying lessons for future regulatory approaches to emerging technologies from the United Kingdom's response to mad cow disease in the early 1990s); Lynn Frewer, *Risk Perception, Social Trust, and Public Participation in Strategic Decisionmaking: Implications for Emerging Technologies*, 28 *Ambio* 569 (1999) (identifying and discussing components of effective policy about risk management in the context of emerging technologies).

[78] Federal Requirements Under the Underground Injection Control (UIC) Program for Carbon Dioxide (CO_2) Geologic (continued...)

---

§ 114 of the Clean Air Act to issue a rule that would require reporting on greenhouse gas emissions from carbon dioxide injection and geologic sequestration.[79] The 111th Congress considered amending the Clean Air Act to broaden the EPA's authority to promulgate similar regulations relating to geologic sequestration.[80]

- *Ocean fertilization*, which may implicate provisions of, *inter alia*, the Marine Protection, Research and Sanctuaries Act of 1972 (MPRSA), 16 U.S.C. § 1431 *et. seq.*, 33 U.S.C. § 1401 *et. seq.*, 33 U.S.C. § 2801 *et. seq.* Title I of the MPRSA prohibits unpermitted ocean dumping by any U.S. vessel or other vessel sailing from a U.S. port in ocean waters under U.S. jurisdiction. Permits may be issued by the EPA if it determines that the dumping will not unreasonably degrade or endanger human health, welfare, the marine environment, ecological systems, or economic potentialities. MPRSA also authorizes NOAA to conduct general research on ocean resources and the EPA to conduct research specifically related to phasing out ocean disposal activities.[81]

- *Stratospheric aerosol injection*, which may implicate the ozone depletion provisions of Title VI of the Clean Air Act Amendments of 1990, 42 U.S.C. § 7401 note. Under those amendments, the Environmental Protection Agency must adjust its phase-out schedules for ozone-depleting substances in accordance with any future changes in Montreal Protocol schedules. The EPA is required to add any substance with an ozone depletion potential (ODP) of 0.2 or greater to the list of Class 1 substances and set a phase-out schedule of no more than seven years. Also, the EPA is required to add any substance that is known or may be reasonably anticipated to harm the stratosphere to the list of Class 2 substances and set a phase-out schedule of no more than ten years.[82]

Moreover, in the absence of federal lawmaking, some states have begun developing their own policies to address particular geoengineering activities.[83]

Potential Roles for Federal Agencies and Other Federally Funded Entities

At this point, federal agencies and other federally funded entities have dedicated minimal efforts and funding to the development and implementation of national geoengineering policies. In

---

(...continued)

Sequestration Wells, 73 Fed. Reg. 43,492 (July 25, 2008). The proposed rule was based on the existing Underground Injection Control (UIC) regulatory framework with modifications to address the unique nature of carbon dioxide injection. Ibid.

[79] Environmental Protection Agency, "Federal Requirements Under the Underground Injection Control (UIC) Program for Carbon Dioxide (CO2) Geologic Sequestration (GS) Wells," 75 *Federal Register* 77230, December 10, 2010.

[80] E.g., American Clean Energy and Security Act of 2009, H.R. 2454, 111th Cong. § 112 (as passed by the House, 2009).

[81] See CRS Report RS20028, *Ocean Dumping Act: A Summary of the Law*, by Claudia Copeland.

[82] See CRS Report RL30853, *Clean Air Act: A Summary of the Act and Its Major Requirements*, by James E. McCarthy et al.

[83] E.g., *Cal. Pub. Resources Code* § 3650(b)(2)(J)(i) (Deering) (supporting research of the ocean's role in carbon sequestration); *Cal. Pub. Util. Code* § 8341(d)(5) (Deering) (permitting utilities to exclude carbon dioxide that they have injected into the Earth from being counted as greenhouse gas emissions); *Tex. Nat. Res. Code* § 91.802 (stating that the Texas Water Code governs anthropogenic carbon dioxide injection wells).

---

testimony before the House Committee on Science and Technology, officials from various interagency bodies coordinating the U.S. response to climate change stated that their offices "(1) have not developed a coordinated research strategy [for geoengineering activities], (2) do not have a position on geoengineering, and (3) do not believe it is necessary to coordinate efforts due to the limited federal investment to date."[84] However, there are numerous potential roles for these entities to play in the development and implementation of national geoengineering policies.

In assessing what agencies should be involved and to what extent, policymakers may consider:

- The advantages or disadvantages of involving multiple agencies and entities;

- The different legislative authorities and areas of expertise that different agencies and entities offer;

- The advantages and disadvantages of relying on independent, executive, and/or legislative bodies; and

- The need to expand or constrict the legislative authority for some federal agencies, so as to give them either more or less jurisdiction over geoengineering activities.

There are, broadly speaking, at least six categories of authorized functions that different federal entities can perform to assist the development and implementation of national policies on new technologies. These categories are (1) conducting research on the science or other aspects of geoengineering, (2) facilitating an exchange of information about geoengineering, (3) funding geoengineering activities, (4) monitoring geoengineering projects and their effects, (5) promulgating regulations, and (6) enforcing regulations. **Table 2** lists selected agencies and entities that currently have the legislative authority to perform various sets of these different functions. These agencies and entities were selected for inclusion in the table because they may assist in the formulation or implementation of future policies on geoengineering, or, in some instances, have already begun to address geoengineering.[85] However, it appears that, to date, no single federal entity is authorized to address the full range of geoengineering technologies.

---

[84] See *Geoengineering III: Domestic and International Research Governance Hearing Before the H. Comm. on Science and Technology*, 111th Cong. (2010) (written testimony by Dr. Frank Rusco, Director of Natural Resources and the Environment, GAO) available at http://democrats.science.house.gov/Media/file/Commdocs/hearings/2010/Full/18mar/Rusco_Testimony.pdf.

[85] For examples of agencies that have already begun to address geoengineering in some fashion, see the discussion in notes for **Table 2**.

Table 2. Six Types of Functions Federal Entities Can Perform and Selected Federal Entities Authorized to Perform Them

Federal Agency or Entity	Conducting Research	Facilitating Information Exchange	Funding	Monitoring Projects or Effects	Promulgating Regulations	Enforcing Regulations
Environmental Protection Agency (EPA)[a]	X	X	X	X	X	X
Department of Energy (DOE)[b]	X		X		X	
Department of Agriculture (USDA)[c]	X	X	X	X	X	X
Army Corps of Engineers (ACE)[d]	X	X		X	X	
National Science Foundation (NSF)[e]	X	X	X			
National Aeronautics and Space Administration (NASA)[f]	X	X	X			
National Oceanic and Atmospheric Administration (NOAA)[g]	X	X	X	X		
United States Global Climate Change Research Program (USGCRP)[h]	X	X	X			

Source: Congressional Research Service.

Notes: This table is intended only as an illustrative list of agencies and entities authorized to perform these types of functions. It is not a comprehensive list of agencies and entities that are addressing or could potentially address geoengineering by performing these functions.

a. The Environmental Protection Agency (EPA) has initiated rulemakings to regulate certain geoengineering activities. In addition, the EPA and other federal bodies have begun funding research and small-scale demonstration projects of certain geoengineering technologies. To date, these efforts have been limited, fragmented, and not coordinated as part of a federal geoengineering strategy.

b. The Department of Energy (DOE) has already sponsored research in both land-based and ocean-based carbon storage, including small-scale demonstration projects of geological sequestration as part of its Regional Carbon Sequestration Partnerships. The agency also funded studies on carbon dioxide air capture techniques. For SRM approaches, DOE, through its Sandia National Laboratories and its Pacific Northwest National Laboratories, has conducted several technology assessments for sulfur aerosol injection and cloud-brightening techniques. However, some suggest that geoengineering technologies not directly related to energy and power generation (i.e., SRM) could remain outside the scope of DOE research and financial support. Moreover, providing grants for research is a different function than providing oversight, regulation, or other governance functions, and the difference could arguably risk conflict-of-interest.

c. The United States Department of Agriculture (USDA) has sponsored research relevant to CDR technologies, including an examination of land-based carbon storage approaches such as biochar through its Agricultural Research Service and carbon sequestration in soil and biomass through its Economic Research Service. Some suggest its continued support for research related to soil, crop, and forest sequestration methods. Beyond research, USDA also provides assistance and incentives to promote deployment of selected technologies, and geoengineering could be added to those. Potential conflict could arise between promotional and regulatory functions.

d. The Department of Defense (DOD) has conducted research on weather modification techniques and other ocean and atmospheric studies. However, many believe that support for geoengineering activities by the defense or intelligence communities could be ill-advised in the international sphere, due to concerns about openness, transparency, and unilateral deployment. The Army Corps of Engineers (ACE), a sub-agency of the DOD with both military and civilian responsibilities, has long-standing responsibilities for projects involving navigable waters, environmental protection, and environmental infrastructure, and, therefore, it was selected for inclusion in the table rather than the DOD.

e. The National Science Foundation (NSF) supports investigator-initiated research through many open and transparent programs and has already funded several projects related to both SRM and CDR approaches, including modeling studies for stratospheric aerosol injection, space-based reflectors, carbon storage in geologic formations, saline aquifers, and biomass. NSF also supports research in the social and behavioral sciences and has conducted studies on the ethical issues of geoengineering. Some suggest its continued support for research related to the science and policy considerations surrounding both CDR and SRM technologies. Again, providing grants is a different function, potentially conflicting with oversight or other governance functions.

f. The National Aeronautics and Space Administration (NASA) has jurisdiction over, *inter alia*, space-related technologies and their use. NASA has previously funded some studies investigating the practicality of using solar shields and reflectors as part of its former Institute for Advanced Concepts program. NASA is a technology development agency, not currently designed for long-term operational or oversight functions.

g. The National Oceanic and Atmospheric Administration (NOAA) appears to have the capacity to model, measure, record, and verify geoengineering technologies as well as to prepare instrumentation, computation, and data analysis. Data NOAA collects about natural climate events also may assist in furthering an understanding of the potential impacts of geoengineering on the ecosystem.

h. The United States Global Climate Change Research Program (USGCRP) conducts ongoing research related to atmospheric circulation, aerosol/cloud interaction, and oceanic chemistry. However, the USGCRP has very limited staff and serves primarily to facilitate cooperation among the various scientific research agencies. It does not have authority for oversight or regulation of technology deployment.

International Cooperation on Geoengineering

When considering forms of international environmental cooperation or oversight, some suggest some form of multilateral agreement that would supplement existing treaties or develop out of ongoing negotiations on other issues of international concern, such as climate change. International agreements have the capacity to codify normative standards for an emerging science on an international level, create institutions for global enforcement and research, and provide a framework under which transparency can be enhanced, development modifications can be made, and future multilateral discussions can occur.

The strengths of international treaties, however, may also be their weakness.[86] Treaties are based on a process that is inherently conservative. Nations often negotiate by adjusting their commitments to a level where they are sure that compliance is technically, economically, socially, and politically feasible. If commitments are perceived as being too high, nations may insert vague language to make the agreement more palatable or simply refuse to join.[87]

Moreover, it can be difficult, particularly when an international situation is new and evolving, to develop international consensus on a set of norms, let alone commitments, given the cultural, political, environmental, and economic diversity of the world's nations. Consequently, the process of developing these norms may be time-intensive and carry the risk of stalemate.[88] Some fear that, given these obstacles, the only "norm" that countries would be willing to agree to at this early stage in the geoengineering science is a moratorium on research and deployment activities. These individuals suggest that those countries who lack the capacity and political incentive to geoengineer may believe there is little to gain from permitting other countries to experiment.

Thirdly, the implementation of international agreements can be difficult to monitor and enforce effectively. On one hand, countries may seek to avoid creating compliance mechanisms and new international institutions on the grounds that they are infringing on a country's sovereignty, and thereby interfering with its ability to experiment with domestic measures that best address local needs and capabilities. On the other, in the absence of such mechanisms, international agreements can be viewed as ineffective if they ultimately fail to change the status quo. Finally, even when international agreements do create new dispute settlement systems, these systems tend to be best equipped to resolve disputes between *countries*, which are considered the principal actors in international law, and not necessarily between one country and one private actor, or between private actors, especially since private actors may shift locations to suit their interests. Consequently, while countries may see advantages to coordinating a global plan for geoengineering activities, local and national laws might still be needed to address aspects of geoengineering that could not be addressed on an international scale.

---

[86] See e.g. David Victor, Kal Raustiala, Eugene Skolnikoff, and Lee Lane. See specifically: David Victor, "On the Regulation of Geoengineering," *Oxford Review of Economic Policy*, Vol. 24, No. 2, 2008, pp. 322-336; as well as comments made in the Royal Society report, op.cit.

[87] Such was the situation in negotiations leading to the 1992 Convention on Biological Diversity, which, as outlined by Victor, op.cit., "contained European-inspired language that was hostile to genetically engineered crops and developing country-inspired language that demanded complicated revenue-sharing for some kinds of germplasm collections. The USA, world leader in these investments, simply refused to join the treaty."

[88] Although, others might argue that opportunities to limit commercial or national activities may decrease as investments are made and economic and political stakes grow larger.

To date, no multilateral treaty has been proposed with the intent of addressing the full spectrum of possible geoengineering activities. However, principles of customary international law and existing international agreements may be implicated by geoengineering research or deployment projects. Governments would likely draw on these principles if they chose to develop a more comprehensive international approach to geoengineering, either by negotiating a new international agreement or expanding upon an existing one. This section will review many of these principles, but, because geoengineering is an umbrella term for a broad array of methods of global climate adjustment, including some that are largely theoretical, it is very likely that particular projects may be affected by international obligations and principles that are not identified in this report.

Principles of Customary International Law

Customary international law results from the general and consistent practice by countries which are followed from a sense of legal obligation.[89] Duties established by customary international law are generally deemed binding on countries that have not persistently objected to it.[90] It can be difficult to determine when a widespread "practice" evolves into a "duty" imposed by customary international law. Nevertheless, under customary international law, countries have a duty not to cause significant transboundary harm.[91] Because geoengineering carries with it the potential for transboundary effects, this duty could be implicated by geoengineering research and/or deployment projects.

International Agreements with Potential Relevance for Geoengineering

In addition to establishing substantive obligations, customary international law also informs the legal significance given by countries to international agreements. As reflected in the Vienna Convention on the Law of Treaties (VCLT), customary international law establishes that signatories of an international agreement must refrain from acts that would defeat the object and purpose of that agreement unless the country makes clear its intent not to ratify the treaty.[92] The VCLT also codifies the customary rule that a treaty may not create rights and obligations for a non-party without its consent.[93] In other words, countries that are not parties to an international agreement may not be bound to adhere to it.

---

[89] *Restatement (Third) of Foreign Relations Law* § 102 (1987).

[90] Ibid. at § 102 n.2.

[91] Ibid. at § 601(1) (stating that a nation is generally obligated to take "such measures as may be necessary, to the extent practicable under the circumstances, to ensure that activities within its jurisdiction or control ... are conducted so as not to cause significant injury to the environment of another state.") Countries are also obligated under international law to take necessary measures to the extent practicable to prevent, reduce, and control pollution that is causing or threatening to cause significant injury to the *marine* environment. Ibid. at § 603(2).

[92] Restatement (Third), *supra* note 89, at § 312(3); VCLT, Art. 18. The United States signed the Vienna Convention on the Law of Treaties (VCLT), but the VCLT has not received the Senate's advice and consent and, consequently, the United States is not a Party to the VCLT. Nevertheless, the United States considers most of the VCLT to constitute customary international law on the law of treaties. See, e.g., Fujitsu Ltd. v. Federal Exp. Corp., 247 F.3d 423 (2d Cir. 2001) ("we rely upon the Vienna Convention here as an authoritative guide to the customary international law of treaties ... [b]ecause the United States recognizes the Vienna Convention as a codification of customary international law ... and [it] acknowledges the Vienna Convention as, in large part, the authoritative guide to current treaty law and practice") (internal citations omitted).

[93] VCLT, Art. 34; *Restatement (Third) of Foreign Relations Law* § 102 (1987), at § 324(1).

The obligations arising from the following treaties and international agreements should be construed in light of these principles of customary international law. The international agreements on climate change are the most likely agreements to have significance for the full spectrum of geoengineering projects because they encourage their parties to implement national policies and mitigation actions to reduce their greenhouse gas emissions. However, these agreements do not currently address geoengineering explicitly. These agreements include:

- *United Nations Framework Convention on Climate Change (UNFCCC).*[94] The UNFCC opened for signature in 1992 and entered into force in 1994. The United States became a party to the UNFCCC in 1992. Under the UNFCCC, parties are required to (1) gather and share information on greenhouse gas (GHG) emissions, national policies, and best practices; (2) launch national strategies for addressing GHG emissions and adapting to expected impacts; and (3) cooperate in preparing for adaptation to the impacts of climate change. Parties are also obligated to cooperate and exchange information on technologies, and potential economic and social consequences of response strategies, as well as to give full consideration to actions to meet the needs and concerns of developing countries that may be adversely affected by, *inter alia*, the implementation of measures to respond to climate change.

- *Kyoto Protocol (the Protocol).*[95] The Kyoto Protocol opened for signature in 1997 and entered into force in 2005. The United States has signed but not become a party to the Kyoto Protocol.[96] The Protocol supplements the UNFCCC by committing its high income parties to legally binding reductions in emissions of greenhouse gases through 2012.

- *Copenhagen Accord (the Accord).*[97] Unlike the Kyoto Protocol and the UNFCCC, the 2009 Copenhagen Accord is a non-binding political agreement. Nevertheless, the United States has indicated its intent to associate with the Copenhagen Accord. The Accord asks Annex 1 Parties of the UNFCCC to set their own individual emissions targets for 2020 and to measure, report, and verify their progress towards these targets pursuant to guidelines adopted by the UNFCCC Conference of the Parties ("COP"),[98] and non-Annex 1 Parties to develop "mitigation actions" for the reduction of GHG emissions, though not emissions *targets*, and measure, report, and verify their implementation of these actions.[99]

- *Convention on Biological Diversity (CBD).*[100] The CBD opened for signature in June 1992 and entered into force in December 1993. The United States has

---

[94] United Nations Framework Convention on Climate Change, May 9, 1992, 1771 U.N.T.S. 107; S. Treaty Doc No. 102-38 *available at* http://unfccc.int/resource/docs/convkp/conveng.pdf.

[95] Kyoto Protocol to the United Nations Framework Convention on Climate Change, Dec. 10, 1997, 37 I.L.M. 22, U.N. Doc FCCC/CP/1997/7/Add.1 *available at* http://unfccc.int/resource/docs/convkp/kpeng.pdf.

[96] As a non-party, the United States is not obligated to comply with the Kyoto Protocol, but, as a signatory, the United States may be obligated to avoid undermining the Kyoto Protocol. For more on this distinction, see CRS Report R41175, *International Agreements on Climate Change: Selected Legal Questions*.

[97] Available at http://unfccc.int/files/meetings/cop_15/application/pdf/cop15_cph_auv.pdf.

[98] Copenhagen Accord, Art. 4.

[99] Ibid. at Art. 5.

[100] United Nations Convention on Biological Diversity, May 22, 1992, 31 I.L.M. 818, available at http://www.cbd.int/ (continued...)

---

signed but has not become a party to the CBD. The key principle of the CBD is that countries have both the sovereign right to exploit their own resources pursuant to their own domestic policies and the responsibility to ensure that activities within their control do not cause damage to the environment of other states or to areas beyond the limits of national jurisdiction.[101] In October 2010, the 10th Conference of the Parties (COP) to the CBD adopted provisions calling for the parties to abstain from geoengineering—including "any technologies that deliberately reduce solar insolation or increase carbon sequestration from the atmosphere on a large scale that may affect biodiversity"—unless the parties have fully considered the risks and impacts of those activities on biodiversity.[102] The COP also affirmed its earlier decision, IX/16C, which acknowledged the work of the London Convention and the London Protocol regarding ocean fertilization and requested that its own Parties ensure that ocean fertilization activities do not take place until either there is adequate scientific basis on which to justify such activities or the activities are small-scale scientific research studies within coastal waters.[103]

In addition to the international climate change agreements, the following international agreements may be relevant to the use of CDR technologies.

- *United Nations Convention on the Law of the Sea (UNCLOS).*[104] UNCLOS opened for signature in December 1982 and entered into force on November 16, 1994. Despite participating in the UNCLOS negotiations, the United States declined to sign the final agreement and has not become a party since then, although it views many provisions of UNCLOS as customary international law that it was already obligated to follow. UNCLOS establishes a legal regime governing activities on, over, and under the world's oceans and defines countries' jurisdictions over, and rights of access to, the oceans and their resources. Article 194 of the UNCLOS imposes a duty on its parties to take, individually or jointly, measures that are necessary to prevent, reduce, and control pollution of the marine environment from any source. The UNCLOS defines pollution as any human-driven introduction of substances or energy into the marine environment that results *or is likely to result* in deleterious effects such as harm to living resources and marine life, hazards to human health, hindrance to marine activities, or impairment of sea water quality.[105] This provision could have significance for a geoengineering project that pollutes the marine environment,

---

(...continued)

convention/convention.shtml.

[101] Ibid. at Art. 3. For example, under Article 8, parties must establish a system and guidelines for the selection of protected areas where special measures need to be taken to conserve biological diversity. Ibid. at Art. 8(a)-(b). They must also regulate, manage, or control the risks associated with the use and release of living modified organisms which are likely to have adverse environmental impacts and must prevent the introduction of, control, or eradicate alien species that threaten ecosystems, habitats, or species. Ibid. at Art. 8(g)-(h).

[102] Conference of the Parties to the Convention on Biological Diversity, Oct.18-19, 2010, *Decision X/33, available at* http://www.cbd.int/climate/doc/cop-10-dec-33-en.pdf.

[103] Ibid.

[104] United Nations Convention on the Law of the Sea, Dec. 10, 1982, 1833 U.N.T.S. 3, 397; 21 I.L.M. 1261 *available at* http://www.un.org/Depts/los/convention_agreements/texts/unclos/unclos_e.pdf.

[105] UNCLOS, Art. 1.1(4).

by land, sea, or air. In addition to, arguably, mandating that a country not engage in that activity, once a geoengineering project resulted in the pollution of the ocean environment, Article 194 would impose a duty on the member country responsible for that pollution to control and limit its spread. Article 192 of the UNCLOS imposes a general obligation on countries to protect and preserve the marine environment. These provisions could be implicated by ocean fertilization and some other geoengineering activities if they have a negative effect on the marine ecosystem.[106] Large-scale ocean fertilization projects could also implicate several UNCLOS provisions, including Article 56, and 238 through 241 on marine scientific research.

- *London Convention on the Prevention of Marine Pollution by Dumping of Wastes and Other Matter (London Convention),*[107] *and London Protocol.*[108] The London Convention was opened for signature in December 1972 and entered into force in August 1975. The London Protocol was agreed to in 1996 as a means of modernizing and eventually replacing the London Convention. The United States became a party to the London Convention in 1974, but it has not become a party to the London Protocol. Contracting parties pledge to take all possible steps to prevent the pollution of the sea by the dumping of substances that are liable to create hazards to human health, harm living resources and marine life, or interfere with other legitimate uses of the sea.[109] However, ocean fertilization and other geoengineering projects that entail the deliberate disposal of substances into the sea might not implicate the London Convention's prohibitions on dumping if they are deemed placed into the sea "for a purpose other than the mere disposal thereof, if not contrary to the aim of the Convention."[110] Nevertheless, the parties have recently considered amendments and resolutions to these agreements to address some geoengineering technologies more explicitly. For example, the 2006 amendments to the London Protocol provide guidance on the means by which *sub-seabed geological sequestration* of carbon dioxide can be conducted, stating that carbon dioxide streams may only be considered for dumping if (1) disposal is into a sub-seabed geological formation; (2) the substances dumped consist overwhelmingly of carbon dioxide; and (3) no other wastes or matter were added to them for the purpose of disposing of them. Two years later, in 2008, the parties adopted Resolution LC-LP.1,[111] agreeing that *ocean fertilization research* activities do not to constitute dumping under the London Convention and Protocol.[112]

---

[106] A thorough review of these "living resources provisions" can be found in CRS Report RL32185, *U.N. Convention on the Law of the Sea: Living Resources Provisions*, by Eugene H. Buck.

[107] London Convention on the Prevention of Marine Pollution by Dumping of Wastes and Other Matter, Dec. 29, 1972, 26 U.S.T. 2403, 11 I.L.M. 1294.

[108] 1996 Protocol to the Convention on the Prevention of Marine Pollution by Dumping of Wastes and Other Matter, Nov. 7, 1996, 36 I.L.M. 7.

[109] London Convention, Art. 1.

[110] Ibid. at Art. 19.1. See also London Protocol, Art. 1.4 (adopting a very similar definition of dumping).

[111] Available at http://www.imo.org/includes/blastDataOnly.asp/data_id%3D24337/LC-LP1%2830%29.pdf.

[112] 2008 Resolution on Ocean Fertilization specifically excludes ocean fertilization *research* from the London Convention and London Protocol's definition of dumping by stating that ocean fertilization is a placement of matter for a purpose other than mere disposal. It urges Contracting Parties to use the "utmost caution and the best available guidance" to evaluate scientific research proposals for ocean fertilization and says that ocean fertilization activities (continued...)

International Cooperation Example—Ocean Fertilization

Ocean fertilization is subject to some level of international governance, but it is unclear who would oversee a worldwide effort to implement ocean fertilization techniques. Most notably, in 2008 a resolution was passed by the Convention on the Prevention of Marine Pollution by Dumping of Wastes and Other Matter (1972, the London Convention) and the 1996 Protocol (London Protocol) to bar ocean fertilization activities other than "legitimate scientific research."[113] The Marine Protection, Research, and Sanctuaries Act of 1972 (MPRSA or Ocean Dumping Act, P.L. 92-532) identifies the U.S. Environmental Protection Agency (EPA) as responsible for enforcing the standards and criteria listed in the London Convention.[114] Also, the United Nations Convention on Biological Diversity (CBD) in 2008 requested member parties to "ensure that ocean fertilization activities do not take place until there is an adequate scientific basis on which to justify such activities."[115] It is possible the United Nations Convention on the Law of the Sea (UNCLOS), which provides an international legal and scientific framework to coordinate international decisionmaking about ocean management, could be applied to the management of ocean fertilization activities. The United States is not a party to UNCLOS.[116]

Finally, in addition to the climate change agreements, the use of SRM technologies may implicate the following international agreements:

- *Convention on the Prohibition of Military or Other Hostile Use of Environmental Modification Techniques (ENMOD or the Convention).*[117] ENMOD opened for signature in May 1977 and entered into force on October 5, 1978. The United States became a party to ENMOD in 1979. ENMOD's aim is to prohibit the military or any other hostile use of environmental modification techniques which have widespread, long-lasting, or severe effects as the means of destruction, damage, or injury to any party. The Convention defines the term "environmental modification techniques" as any technique for changing—through the deliberate manipulation of natural processes—the dynamics, composition, or structure of the Earth, including its biota, lithosphere, hydrosphere, atmosphere, or outer space.[118] This definition may encompass certain geoengineering activities. However, the Convention further states that the provisions of the Convention do not hinder the use of environmental modification techniques for peaceful purposes, such that parties to the Convention may undertake to facilitate, and

---

(...continued)

other than research should not be allowed given the present state of knowledge.

[113] Additional details regarding the stipulations of the London Convention are provided on page 27 of this report. The United States became a party to the London Convention in 1974, but it has not become a party to the London Protocol.

[114] For more information on the Ocean Dumping Act, see CRS Report RS20028, *Ocean Dumping Act: A Summary of the Law*, by Claudia Copeland.

[115] "COP 9 Decision IX/16," Biodiversity and climate change, http://www.cbd.int/decision/cop/?id=11659. Additional details regarding the Convention on Biological Diversity is available on page 32 of this report. The United States has signed but has not become a party to the CBD.

[116] UNCLOS is currently before the Senate Foreign Relations Committee and awaiting the Senate's advice and consent on the question of U.S. accession.

[117] Convention on the Prohibition of Military or Other Hostile Use of Environmental Modification Techniques, May 18, 1977, 31 U.S.T. 333, 16 I.L.M 88, available at http://www.un-documents.net/enmod.htm.

[118] Ibid. at Art. II.

---

have the right to participate in, the fullest possible exchange of scientific and technological information on the use of environmental modification techniques for peaceful purposes.[119]

- *Convention on Long-Range Transboundary Air Pollution (CLRTAP or the Convention).*[120] CLRTAP opened for signature in 1979 and entered into force on March 16, 1983. The United States became a party to the Convention in 1979. Contracting parties commit themselves to limiting, and to gradually preventing and reducing their discharges of air pollutants and thus to combating the resulting transboundary pollution. Long-range transboundary air pollution is defined by the Convention as the human introduction of substances or energy into the air which have deleterious effects on human health, the environment, or material property in another country, and for which the contribution of individual emission sources or groups of sources cannot be distinguished.[121] It is uncertain which geoengineering activities CLRTAP would regulate, or how such regulation would be implemented.

- *Vienna Convention for the Protection of the Ozone Layer (the Convention).*[122] The Vienna Convention for the Protection of the Ozone Layer was opened for signature in 1985 and entered into force in 1988. The Convention, along with the Montreal Protocol on Substances that Deplete Ozone Layer, which was opened for signature in 1987 and entered into force in 1989, aims to control the production and consumption of the most commercially and environmentally significant ozone-depleting substances.[123] The United States became a party to the Vienna Convention in 1986. It is also a party to the Montreal Protocol (since 1988) and has agreed to all four amendments to the Protocol. Parties to these agreements, within their capabilities, are expected to (1) cooperate to better understand and assess the effects of human activities on the ozone layer and the effects of the modification of the ozone layer; (2) adopt appropriate measures and cooperate in harmonizing appropriate policies to control the activities that are causing the modification of the ozone layer; (3) cooperate in formulating measures for the implementation of the Convention; and (4) cooperate with competent international bodies to implement the Convention.[124] Certain stratospheric aerosol injection technologies for geoengineering, to the extent that they pose a risk to the ozone layer, have the potential to implicate many of the phase-out provisions of the Convention and the protocols.

---

[119] Ibid. at Art. III.

[120] Convention on Long-Range Transboundary Air Pollution, Nov. 13, 1979, 34 U.S.T. 3043, 18 I.L.M. 1442, available at http://www.unece.org/env/lrtap/full%20text/1979.CLRTAP.e.pdf.

[121] Ibid. at Art. 1.a-b.

[122] Vienna Convention for the Protection of the Ozone Layer, Mar. 22, 1985, T.I.A.S. No. 11,097, 13 I.L.M. 1529, available at http://www.unep.ch/ozone/pdfs/viennaconvention2002.pdf.

[123] Ibid. at Art. 2.1.

[124] Ibid. at Art. 2.2.

The Relevance and Functions of Various International Bodies

At this point, no international organization has a direct mandate to address the full spectrum of possible geoengineering activities.[125] It is possible, however, that existing institutions could fit this purpose if their charters were modified and expanded. Assessing what kinds of institutions would be best suited for this enterprise is difficult, given how little is currently understood about the technical, economic, social, and political components of the technologies. Consequently, there is debate over the ideal structure and framework of an international institution that should have responsibility for geoengineering activities or policies.

There are many factors to consider before deciding whether to create a new international body to address geoengineering or to grant jurisdiction over geoengineering to an existing international body or group thereof. Among these factors are:

- The functions the international body should perform (see **Table 2**);

- The level of membership and inclusiveness the international body should have;

- The level of resources and experience on which the international body should be able to draw;

- The appropriate subject-area jurisdiction, or jurisdictions, that the international body should have; and

- The voting rules that will best enable the international body to make careful inclusive decisions but still respond with appropriate speed to new issues.

Some observers suggest that, because engineering the climate system is a global activity with transboundary effects, only a multilateral body is appropriate for addressing it. This kind of body could be the United Nations, a specialized body or agency contained within it, such as the United Nations Environment Programme (UNEP) or International Maritime Organization (IMO), or an environmental convention secretariat associated with it, such as the secretariat for the United Nations Convention on Climate Change (UNFCCC). Among the advantages of involving this kind of international institution are that it typically has (1) a truly international reach, (2) access to a large budget and resources, (3) legitimacy and leverage with different countries and stakeholders, and (4) experience in handling and developing consensus around controversial issues on an international scale. However, while a multilateral body may assist in bringing transparency, inclusiveness, and equitability to geoengineering oversight, the U.N. process is slow and complicated by design and may fail to provide a sufficiently rapid response to geoengineering activities.[126] Furthermore, some worry that folding geoengineering into the jurisdiction of a pre-existing multilateral institution might result in mission creep and conflicts of interest. These

---

[125] However, in the absence of any official international entity charged with addressing geoengineering activities, three organizations founded the Solar Radiation Management Governance Initiative (SRMGI). *About SRMGI*, http://www.srmgi.org (last visited Jan. 7, 2011). The three founding partners of SRMGI are (1) the Royal Society, which is funded in part by the United Kingdom; (2) the Academy of Sciences for the Developing World, an autonomous international organization funded primarily by Italy and administered by the United Nations; and (3) the Environmental Defense Fund, a U.S. nonprofit focused on environmental protection and advocacy. *Id.* SRMGI plans to release a set of recommendations for the governance of geoengineering—particularly, SRM—research in the spring of 2011. *Id.* SRMGI states that will involve a wide variety of stakeholders in the development of its recommendations, including representatives of governments of the developed and developing countries. *Id.*

[126] See Virgoe's written and oral testimony to the U.K. House of Commons, op. cit. The testimony also includes the critique that follows.

---

critics suggest that the institution might be biased against geoengineering because of its primary mission or, alternately, that it might be forced to allocate less of its time and resources to achieving that primary mission.

Another type of international organization that could be involved in geoengineering is that of plurilateral *ad hoc* bodies, which are frequently groups of countries with shared characteristics that might facilitate dialogue and coordination on particular issues. For example, members of groups like the G-20 or Major Economies Forum (MEF) may share similar economic situations, technical abilities, or climate concerns that would enable these groups to coordinate research agendas and facilitate the exchange of information. But, because these groups have limited membership, their involvement might undermine the legitimacy of their response to geoengineering. The countries who are excluded from participating from the plurilateral organization, and therefore from formulating its response to geoengineering, may demand adequate voice and representation. Moreover, the same issues of mission creep and conflict of interest might plague the use of existing plurilateral groups just as they would the use of existing multilateral institutions.

A third type of international organization that could be involved is that of the intergovernmental organization, which typically acts a policy advisor to its member countries. The International Energy Association (IEA), for example, has a mandate to coordinate government measures affecting energy security, economic development, and environmental protection. These measures include those related to carbon capture and sequestration, and the IEA recently published a report assessing its members' progress towards implementing CCS projects and setting recommendations for next steps.

A fourth type of international organization that could be involved is that of the international nongovernmental organization, which is exemplified by the International Organization for Standardization (ISO). The ISO is a network of the national standards institutes of 163 countries, some of which are affiliated with their national governments and others of which are more closely associated with the private sector. The ISO develops an international set of standards in response to a clearly established need by a particular sector or group of stakeholders. Most ISO members have some form of public review procedures so that outside feedback can be incorporated into the draft standard. For that standard to then be accepted as an ISO International Standard, it must be approved by at least two-thirds of the ISO national members. To date, the ISO has not developed a comprehensive set of standards addressing geoengineering activities. However, it has created standards in other potentially related areas including air quality, water quality, and the quantification and reporting of greenhouse gas emissions.

Finally, international research consortia represent a fifth type of potentially relevant international bodies. Research consortia are generally well equipped to (1) set scientific research priorities at the initial stages of an emerging technology; (2) explore and evaluate the feasibility, benefits, risks, and opportunities presented by an emerging technology; (3) coordinate existing research, identify new research agendas, and develop effective and objective assessment frameworks to inform the initial stages of regulation; (4) collaborate with the scientific, policy, commercial, regulatory, and nongovernmental communities to provide independent oversight of evolving regulatory issues for an emerging technology; and (5) formulate, develop, and socialize an international and voluntary code of practice to govern research in an emerging technology to provide guidance and transparency for the public, private, and commercial sectors. Accordingly, loosely coordinated international consortia could support cooperative geoengineering research through transparent and informal consultations on risk assessment, acceptability, and oversight.

These collaborations could engage a broad group of experts and stakeholders, from scientists to public policy-makers to civil society and explore the safest and most effective forms of geoengineering while building a community of responsible researchers. International policy norms on geoengineering could then be built from the bottom up, as knowledge and experience regarding geoengineering technologies continued to develop. Interactive links between emerging governance and ongoing scientific and technical research could be the core of this approach.[127] Observers point to similar international collaborations where the science has had potentially hazardous side effects such as the European Organization for Nuclear Research (CERN) and the Human Genome Project.[128] Currently, however, no collaborative mechanism is applicable to geoengineering.

Notably, any international body granted jurisdiction over geoengineering will likely lack the authority to fully regulate or enforce its members' compliance with the terms of the body's charter or the underlying international agreement. Even international bodies with dispute settlement bodies or mechanisms in place depend, ultimately, on their members' cooperation to conform their measures and actions with the terms of either the international agreement in question or any decision reached pursuant to the dispute settlement process. For example, the International Court of Justice (ICJ) handles disputes between Members of the United Nations and other countries that have either become parties to the Statute of the Court or otherwise accepted the ICJ's jurisdiction. However, if a party to an ICJ dispute fails to act in accordance with the ICJ's judgment, the other party may present the matter to the U.N. Security Council, which *may* "if it deems necessary, make recommendations or decide upon measures to be taken to give effect to the judgment."[129] Similarly, the World Trade Organization has a dispute settlement process for disputes involving allegations of a Member's non-compliance with the provisions of the General Agreement on Tariffs and Trade (GATT) or one of the WTO agreements. However, it is ultimately up to the WTO Dispute Settlement Body, which is composed of representatives of all WTO Members, to authorize the complaining Member to retaliate (generally by raising tariffs against) the defending Member for non-compliance with a WTO panel or Appellate Body decision.[130] Consequently, while the decisions rendered by the ICJ and the WTO panels and Appellate Body generally persuade countries to comply, international institutions must ultimately rely on international negotiations and diplomacy to ensure universal compliance with the underlying agreements.

Conclusion

Geoengineering is an emerging field that, like other areas of scientific innovation, requires careful deliberation by policymakers, and possibly, the development or amendment of international agreements, federal laws, or federal regulations. Currently, many geoengineering technologies are at the conceptual and research stages, and their effectiveness at reducing global temperatures has

---

[127] As proposed by David Keith, Edward Parson, and M. Granger Morgan, "Opinion: Research on Global Sun Block Needed Now," *Nature,* Vol. 463, No. 28, January 2010, pgs. 426-7.

[128] See Victor, op. cit.

[129] U.N. Charter, Art. 94(2).

[130] For more on the WTO dispute settlement system, read CRS Report RS20088, *Dispute Settlement in the World Trade Organization (WTO): An Overview*, by Daniel T. Shedd, Brandon J. Murrill, and Jane M. Smith. Notably, the WTO Dispute Settlement Body applies a reverse consensus voting rule for the authorization of retaliation: the DSB will adopt a panel or Appellate Body decision authorizing retaliation unless, by consensus, it decides *not* to adopt it.

---

yet to be proven. Very few studies have been published documenting the cost, environmental effects, socio-political impacts, and legal implications of geoengineering. Nevertheless, if geoengineering technologies are deployed, they are expected to have the potential to cause significant transboundary effects.

Some foreign governments and private entities have expressed an interest in pursuing geoengineering projects, largely out of concern over the slow progress of greenhouse gas reductions under the international climate change agreements, the possible existence of climate "tipping points," and the apparent political or economic obstacles to pursuing aggressive domestic greenhouse gas mitigation strategies. However, in the United States, there is limited federal involvement in, or oversight of, geoengineering. Consequently, to the extent that some federal agencies and U.S. states have begun addressing geoengineering projects, they are doing so in a largely piecemeal fashion.

If the U.S. government opts to address geoengineering at the federal level, there are several approaches that are immediately apparent. First, it may continue to leave geoengineering policy development in the hands of federal agencies and states. Second, it might impose a temporary or permanent moratorium on geoengineering, or on particular geoengineering technologies, out of concern that its risks outweigh its benefits. Third, it might develop a national policy on geoengineering by authoring or amending laws. Fourth, it could work with the international community to craft an international approach to geoengineering by writing or amending international agreements. That the government can play a substantial role in the development of new technologies has been manifested in such areas as nanotechnology, nuclear science, and genetic engineering.

Author Contact Information

Kelsi Bracmort
Specialist in Agricultural Conservation and Natural Resources Policy
kbracmort@crs.loc.gov, 7-7283

Richard K. Lattanzio
Analyst in Environmental Policy
rlattanzio@crs.loc.gov, 7-1754

www.ingramcontent.com/pod-product-compliance
Lightning Source LLC
Chambersburg PA
CBHW081237170526

45165CB00009B/3086